Recent Titles in This Series

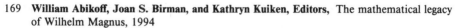

(Continued in the back of this publication)

Foundational Aspects of "Non"standard Mathematics

CONTEMPORARY MATHEMATICS

176

Foundational Aspects of "Non"standard Mathematics

David Ballard

American Mathematical Society
Providence, Rhode Island

*Sci
QA
299.82
B35
1994*

1991 *Mathematics Subject Classification*. Primary 03H05; Secondary 54D99, 03C62.

ABSTRACT. Early in the development of Nonstandard Analysis Luxemburg noted natural topologies (the "S-topologies") to exist on the internal part of a Robinson enlargement. In this work these are generalized and used to give new, topological foundations for Nonstandard Mathematics. The resulting topological methods are then applied to construct models (implying conservativity over ZFC) of the nonstandard set theories proposed by Nelson, Hrbáček and Kawai. A simple yet nontrivial extension of a nonstandard set theory of Fletcher's is then described and proposed as a prototype of the "ultimate" vehicle for Nonstandard Mathematics. Although the mathematical environment it presents is radically relativistic, it is never the less shown to be "safe" (conservative over ZFC) for practitioners.

Library of Congress Cataloging-in-Publication Data

Ballard, David, 1941–
    Foundational aspects of "non"standard mathematics / David Ballard.
        p.   cm. — (Contemporary mathematics, ISSN 0271-4132; v. 176)
    Includes bibliographical references and index.
    ISBN 0-8218-0293-3 (acid-free)
    1. Nonstandard mathematical analysis. I. Title. II. Title: Foundational aspects of nonstandard mathematics. III. Series: Contemporary mathematics (American Mathematical Society); v. 176.
QA299.82.B35   1994
511′.2—dc20

                                                                                94-30811
                                                                                CIP

Contents

Introduction

This entire work is the result of an innocent question. At the ICM '86 in Berkeley I happened to listen to Bill Davidon deliver a paper [6] on a proposed "non"standard set theory in which ordinary sets, if infinite, would include shadow elements or "guests". Most such set theories come with an axiom of "saturation" or "idealization" but in Davidon's system the axiom was named "compactness". For me this evoked the question: *what is the topology?*

In the following months of discussion I had with Davidon a preliminary answer emerged: the "internal" part of a "non"standard system turns out to carry a natural topology with quite distinctive features and from which the basic notions of "non"standard analysis can be re-expressed in purely topological terms. I was also curious to discover that this topology had been noted some twenty years earlier by Luxemburg [16] and called by him the "S-topology". At the time "non"standard analysis was undergoing rapid development and perhaps the scramble and haste for results that was typical then can explain why such an innocent observation — this "S-topology" — remained unexamined for so long.

Being a newcomer to the arena I had a rather different take. Some fields of mathematics seem to resonate deeper than others. They seem to tap into a more fundamental perspective of things. To me, point set topology is clearly such a field and if an entire specialty, such as "non"standard analysis, can be re-expressed in its terms, then something fundamental is obviously being touched and begs investigation. Intrigued, I undertook the investigation.

This work is the result. As "non"standard analysis grows out of and borrows concepts from model theory, and as indeed the topological perspective I develop for "non"standard analysis spreads well into sections of model theory proper, this work sits at the cross roads of *three* fields of mathematics, namely, point set topology, model theory, and "non"standard analysis.

I am not a specialist in any of the three mathematical fields just mentioned. Rather, I am a *visitor* and the results developed in this text are typical of the discoveries that a visitor might make, namely, of a middle level of technicality, open ended in nature, and yet overlooked by the arena's long term residents. I think it will be helpful to keep this simple fact in mind while reading this work.

To be accessible to a broad audience of mathematicians, I begin with a background review in Part 1 of the concepts needed from point set topology, model theory, and "non"standard analysis. The review is written with the eye of the visitor. Connections are dwelt on and, where appropriate, new results are included which further illustrate the point. Indeed, during the discussion of model theory I will show that several of its key concepts, in particular saturation, are naturally re-expressed in terms of point set topology. This suggests a generalization of model theory proper, which I proceed to sketch briefly. Called "point set model theory", this generalized theory applies more smoothly to "non"standard mathematics than does its traditional cousin. In this theory, one is able to isolate certain structures (termed "internal domains") which model the internal parts of existing "non"standard theories.

In Part 2, the point set topology relevant for "non"standard theories is introduced. Characteristic spaces ("CL spaces") are first isolated and explored. From these, a topological object is defined which is shown to exactly match those internal domains which are "local". Appropriate morphisms between local internal domains are also captured in topological terms. Finally, it is shown how direct limits of these local internal domains result in canonically defined topological objects which exactly match the class of all internal domains. The appropriate morphisms between internal domains also reduce to topology. In this manner, the internal parts of "non"standard theories are seen to form a chapter in classical point set topology.

Armed with this insight, a major re-evaluation of "non"standard theories is undertaken in Part 3. After Robinson's initial formulation of "enlargements" in the 1960's (see [20]), there have been, starting in the mid 1970's, a series of proposed "non"standard set theories which sought to make "non"standard mathematics "global". Part 3 begins with a review of enough of standard set theory (ZFC and its variants) to make discussion of these intelligible. After a unified description of the main "non"standard set theories presented in the literature (Nelson's IST [18], Hrbáček's NS_1 and NS_2 [11] and Kawai's NST [14] — with Fletcher's $SNST$ [9] reserved for later discussion), I construct models of each theory relying on the topological methods of Part 2. The manner in which the constructions are done is uniform and automatically implies that each "non"standard theory is "conservative" (doesn't imply new "standard" theorems). At this point, we pause for a critical review of all such "non"standard set theories, with Fletcher's $SNST$ now included in the discussion. It is an interesting and stubborn fact that none of these theories gave a freewheeling vehicle capable of serious, global "non"standard mathematical flight. In a sense, we pour through the wreckage, seeking lessons and reasons. Mindful of certain basic limitations for "non"standard set theories early noted by Hrbáček in [11], I develop a shopping list of desired features which the "true" "non"standard set theory ought to possess. Making a minor yet nontrivial modification of Fletcher's $SNST$, I then propose such a set theory, to be called EST. As done for the previous theories, I construct a model of EST in a manner which shows its conservativity over ZFC. Part 3 ends with a frank, open ended discussion of EST's potential usefulness to the greater mathematical community.

Among potential readers of this work it is the model theorist who may experience a certain skepticism. My viewpoint is decidedly topological and, wherever possible, model theoretic concepts are rephrased within topology. Although novel viewpoints tend to offer new perspective, a legitimate question of substance remains. Is a rephrased model theory actually new model theory? In my view, the topological methods I develop in this work do represent new model theory. Specifically, the main result of Part 2, as embodied in Theorem 7.1, is new model theory which, at best, is awkward to even describe in traditional model theoretic terms. Also, the use of topology in the proof of Theorem 13.1, concerning the conservativity of EST over ZFC, allows one to avoid the traditional use of ultralimits. It is difficult for me to imagine a proof of this theorem, using ultralimits, which anyone could possibly read.

Having said this, I want to also point out that I have endeavored to keep the set theoretic portion of this work, Part 3, in conventional terms as much as possible. All topology (with the exception of the statement of Theorem 11.2 which requires it) is confined to the body of proofs. In particular, a model theorist or set theorist should have no difficulty following all the discussion, definitions, statements of results, and even proofs (excepting that of Theorem 13.1) which concern the system EST.

I want to warmly acknowledge the role that Bill Davidon played in the creation of this work. He shared in the original discovery with me in the Fall of '86 and his companionship in the years that followed during which details were being worked out was crucial. Without him I could have never completed the work. I also deeply appreciate the several detailed constructive readings that Karel Hrbáček generously gave the manuscript in its various drafts. His many suggested revisions were extremely helpful. It was the sage advice of Ernst Snapper that led me to find an Editor sympathetic to this work. One's simple acts can be so helpful to another. Chris Colebank provided crucial technical assistence with the software used in the final electronic submission of the work. Portions of the last draft were kindly reviewed by Norm Feldman and Larry Green. I am thankful to the members of the Mathematics Department of Sonoma State University for making their facilities available to me.

And it is my real pleasure to dedicate this work to my grandchildren Alexandra Bendayan-Hood, Colin Randall Hood, Kendrah Rose Gilman, and Larissa Renée Gilman.

DEPARTMENT OF MATHEMATICS, SONOMA STATE UNIVERSITY, ROHNERT PARK, CALIFORNIA 94928
    Current address: Department of Mathematics, Sonoma State University, Rohnert Park, California 94928
    E-mail address: David.Ballard@Sonoma.edu

Part 1

Preliminaries

CHAPTER 1

Point Set Topology

Despite its humble appearance, point set topology is a rich vehicle for mathematical insight. Indeed, it is to be counted among one of the fundamental fields of mathematics. There are many ways to define a topological space, but for mathematical convenience the following is traditional. A *topological space* is taken to be a set X of points with a specified collection $\mathcal{T}$ of its subsets satisfing the axioms

$$\emptyset, X \in \mathcal{T},$$
$\mathcal{T}$ is closed under finite intersections,
$\mathcal{T}$ is closed under arbitrary unions.

The subsets $U \in \mathcal{T}$ are are called *open* and their complements $C = X - U$ are called *closed*. A subset N in X which contains an open set U which contains a point $p \in X$ is called a *neighborhood* of p.

Geometrically, we can think of the fact of a point $p \in X$ possessing a neighborhood N which is disjoint from a subset Y as being evidence that, not only is p not an element of Y, but that p does not even "touch" Y, that indeed, with sufficient magnification, p can be viewed as being "positively away" from the region that Y makes up. Indeed, if we define the meaning of p *touches* Y to be that all its neighborhoods have a nonempty intersection with Y (this includes the case when $p \in Y$), then a number of geometric notions are naturally defined on a topological space. The *boundary* of Y (written $Bd(Y)$) is the collection of all points p which touch both Y and its outside, i.e., the complement $Y' = X - Y$. The *interior* of Y (written Y°) consists of all elements $p \in Y$ which are so far inside Y that they do not touch the outside $Y' = X - Y$. Expressed otherwise, $Y^\circ = Y - Bd(Y)$. The *closure* of Y (written $\overline{Y}$) consists of all its members along with any other points which touch it, that is, $\overline{Y} = Y \cup Bd(Y)$. Equivalently, the closure $\overline{Y}$ is the intersection of all closed sets C which contain Y. A set Y turns out to be *closed* if it coincides with its closure, that is, if $Bd(Y) \subseteq Y$. Dually, the set Y turns out to be open if it coincides with its interior, i.e., if $Y \cap Bd(Y) = \phi$. An element $p \in Y$ is *isolated* (in Y) if it does not touch what remains of Y when p is removed, that is, if p does not touch $Y - \{p\}$. If instead $p \in Y$ does touch the rest of Y, then p is said to be a *limit point* of Y. If $Y \subseteq Y'$ are subsets of

7

X and every point of Y' touches Y (so $Y' \subseteq \overline{Y}$), then Y is said to be *dense* in Y'. If X' is a second topological space, then a mapping $f : X \to X'$ is *continuous* if it preserves touching, i.e., if p touches Y, then $f(p)$ will touch $f[Y]$ (here $f[Y] = \{f(p) : p \in Y\}$ is the f-image of the set Y). Equivalently, $f : X \to X'$ is continuous if the inverse image $f^{-1}[U'] = \{p : f(p) \in U'\}$ of any open set U' in X' is open in X. If $X \subseteq X'$ is a subset and the inclusion map f is not only continuous but the reverse implication holds (p touches Y whenever $f(p)$ touches $f[Y]$), then X is said to be a *subspace* of X' or to have the *subspace topology* induced by X'. This is the same as saying the typical open set U in X has the form $U = X \cap U'$ for some open set U' in X'. If $f : X \to X'$ is continuous, one to one, and onto and if the inverse map f^{-1} is also continuous, then f is said to be a *homeomorphism*, and X and X' are said to be *homeomorphic* as topological spaces. We shall also call a continuous map $f : X \to X'$ an *immersion* if it is a homeomorphism $X \to f[X]$, where $f[X]$ is given its X'-subspace topology.

Given a topological space X with collection of open set $\mathcal{T}$, we can speak of a subcollection $\mathcal{T}' \subseteq \mathcal{T}$ as generating the topology, or being a *basis* for X, if every open set in X is a union of elements of $\mathcal{T}'$. If the finite intersections of elements of $\mathcal{T}'$ form such a basis, then $\mathcal{T}'$ is called a *subbasis* for X. If $\mathcal{T}' \subseteq \mathcal{T}$ and both give topologies on X, then, of the two, $\mathcal{T}'$ is the *coarser* topology, while $\mathcal{T}$ is the *finer*. Clearly, the finer the topology the sharper the focus and the less point set touching is perceived. The finest topology on X (the *discrete* topology) is the one where $\mathcal{T}$ consists of all subsets of X. In this topology the only points p touching a set Y are its very elements. If $\mathcal{T}'$ and $\mathcal{T}$ give topologies on sets X and X', respectively, then the *product topology* on $X \times X'$ is the coarsest topology for which both projection maps $X \times X' \to X$ and $X \times X' \to X'$ are continuous. Subsets of the form $U \times U'$, where $U \in \mathcal{T}$ and $U' \in \mathcal{T}'$, form an open basis for this topology. Such product topologies naturally generalize to cartesian products of arbitrary families of topological spaces.

A common situation in the topological spaces we shall meet later is one where X and all its finitary powers X^n have given topologies which are not necessarily the product topologies which would be induced from that on X, but yet are such that all projection maps $X^n \to X^m (m < n)$ are continuous. From this, arbitrary powers X^I of X can receive canonical topologies as follows. For any finite choice $i_1, i_2, \dots, i_n$ of distinct elements of I, consider the "projection" $X^I \to X^n$ which sends the element $f \in X^I$ to the element $(f(i_1), f(i_2), \dots, f(i_n)) \in X^n$. We then give X^I the coarsest topology which leaves all such "projections" continuous. If the original topologies on the X^n are "symmetric" (all maps $X^n \to X^n$ obtained by permuting coordinates are continuous), then the induced topology on X^I, with $I = n = \{0, 1, \dots, n-1\}$, coincides with the original. A useful observation concerning these topologies is the following:

PROPOSITION 1.1. *Assume the the finitary powers X^n of a set X are topologized so that all projection maps $X^n \to X^m (m < n)$ are continuous. Let I be an* *arbitrary index set and give X^I its induced topology. Suppose $I = \bigcup_{i \in \Gamma} I_i$ where the family of sets $\{I_i : i \in \Gamma\}$ is closed under finite unions. For each $i \in \Gamma$ let $\pi_i : X^I \to X^{I_i}$ be the projection sending $f \in X^I$ to $f_{|I_i} \in X^{I_i}$. Then for any subset $D \subset X^I$, $\overline{D} = \bigcap_{i \in \Gamma} \pi_i^{-1}[\overline{\pi_i[D]}]$.*

PROOF. Since the projections involved are continuous, the inverse images of these closed sets are all closed and contain D and, therefore, so does their intersection. Thus, we have $\overline{D} \subseteq \bigcap_{i \in \Gamma} \pi_i^{-1}[\overline{\pi_i[D]}]$. Conversely, suppose $f \in X^I$ does not lie in $\overline{D}$. Pick an open $f \in U$ such that $U \cap D = \emptyset$. By definition of the topology on X^I, we can assume there exists finite $I' \subseteq I$, and open $U' \subseteq X^{I'}$, such that if $\pi' : X^I \to X^{I'}$ is the corresponding projection, then $U = \pi'^{-1}[U']$. Since I' is finite, it must be contained in some I_i and, since π' "factors through" π_i, there will exist open $U_i \subseteq X^{I_i}$, so that $U = \pi_i^{-1}[U_i]$. But then $f_{|I_i} \notin \overline{\pi_i[D]}$ and hence, $f \notin \pi_i^{-1}[\overline{\pi_i[D]}]$. This proves the reverse inclusion. □

Thinking as a logician, an alternate view of topological spaces is possible. Heuristically, one can think of a set Y within a topological space X as a "property" which the elements p of X may or may not possess. However, the true "extension" of this property should not be confined to Y, but should really extend to all of its closure $\overline{Y}$. Pursuing this analogy, we can think of describing an element $p \in X$ by listing its "properties", i.e., by naming the closed sets Y in X to which it belongs. Conversely, one can use a collection $\{Y_i : i \in \Gamma\}$ of closed sets in X as a description of a potential element, and inquire if such an element exists in X. If the intersection of all the Y_i is empty, then clearly none does. However, when every possible finite subdescription (i.e., a $\{Y_i : i \in \Gamma_0\}$ where $\Gamma_0 \subseteq \Gamma$ is finite) does manage to describe an element of X, then one can certainly feel that the main collection $\{Y_i : i \in \Gamma\}$ *ought* to as well. When this happens in every case, the space X is said to be *compact*. This idea can be defined more rigorously as follows: A collection $\{Y_i : i \in \Gamma\}$ of sets is said to satisfy the *finite intersection property* (f.i.p. for short) if the intersection of every finite subcollection is nonempty. The space X is compact if every collection $\{Y_i : i \in \Gamma\}$ of its closed sets which satisfies f.i.p. has a nonempty intersection.

The notion of compactness is quite fundamental. In disguised form and under numerous names it crops up in many places. The "completeness" of the real numbers is equivalent to the compactness of each of its closed finite intervals. We will see in model theory that the notion of "saturation" is essentially a phenomenon of compactness. This also occurs in "non"standard mathematics where some authors have named it "idealization".

In terms of open sets, compactness can be defined as follows: A subcollection $\mathcal{T}' \subseteq \mathcal{T}$ of open sets is an *open cover* if the union of its elements is all of X. The space X is compact if every one of its open covers $\mathcal{T}'$ contains a subcover $\mathcal{T}'_0 \subseteq \mathcal{T}'$ which is finite.

From a logician's point of view, the "properties" in a topological space will seem to fit a rather rudimentary "logic", as compared to the classical two valued

logic usual in ordinary mathematical discourse. Finite disjunctions (use of the
connective "or") exist, since the union of finitely many closed sets is closed.
Even infinite conjunctions (use of the connective "and") exist, as the intersection
of arbitrarily many closed sets is also closed. However, negations (use of the
connective "not") generally fail. The property "not Y" may not exist on X in
the classical two valued sense: it would have to correspond to $\overline{Y'}$ ($Y' = X - Y$),
and this may well have a nonempty intersection with $\overline{Y}$. A classical "not Y"
will occur precisely when $Bd(Y) = Bd(Y') = \emptyset$. This happens when the set Y is
both open *and* closed. Such sets are called *clopen*. Clearly a topological space X
which embodies a lot of classical logic is one which has many clopen subsets. We
can take this to mean that X has an open basis made up of its clopen subsets.
Such spaces have traditionally been called *zero dimensional*. We will largely be
concerned with zero dimensional spaces in this work.

As a further exercise with these concepts, and for its use later, we shall include
here the proof of another simple result. A continuous mapping $f : X \to X'$
between topological spaces may have further properties. The mapping f is *open*
if it preserves open sets ($U \subseteq X$ open implies $f[U] \subseteq X'$ also open). Similarly it is
closed if it preserves closed sets. It preserves *closures of points* if $f[\overline{\{p\}}] = \overline{\{f(p)\}}$
for every $p \in X$.

PROPOSITION 1.2. *Let $f : X \to X'$ be a continuous map from a compact
space X to a zero dimensional space X'. If f preserves closures of points, then
f is a closed mapping.*

PROOF. Let $C \subseteq X$ be any closed set and let $p' \in X'$ be any point not in
$f[C]$. We first observe that since f preserves closures of points and $p' \notin f[C]$ we
must have $\overline{\{p'\}} \cap f[C] = \emptyset$. For otherwise we would have some $p \in C$ for which
$f(p) \in \overline{\{p'\}} \cap f[C]$ and hence,

$$\overline{\{p\}} \subseteq C \text{ (since } C \text{ is closed)} \Rightarrow$$
$$\overline{\{f(p)\}} = f(\overline{\{p\}}) \subseteq f[C] \text{ (since } f \text{ preserves closures of points)}$$
$$\Rightarrow p' \notin \overline{\{f(p)\}} \text{ (since } p' \notin f[C]).$$

But X' is zero dimensional, so we would then have clopen $p' \in U'$ such that
$U' \cap \overline{\{f(p)\}} = \emptyset$ and, since U' is clopen, this would impy $\overline{\{f(p)\}}$ and $\overline{\{p'\}}$ are
disjoint, which contradicts the fact that $f(p) \in \overline{\{p'\}}$. Thus $\overline{\{p'\}} \cap f[C] = \emptyset$ and
$f[C]$ can be covered by clopen subsets none of which contain the point p'. But
C is a closed subset of a compact space, hence is compact as a subspace, and
since such sets are always preserved under continuous maps, we must have that
$f[C]$ is a compact subspace of X'. Thus the clopen cover of $f[C]$ can be replaced
by a finite subcover, and since finite unions of clopen sets are still clopen, we
derive a single clopen set in X' which contains all of $f[C]$, but not the point
p'. Therefore, the complement of this clopen is an open set containing p' which
is disjoint from $f[C]$. We see now that p' is not in the closure of $f[C]$. More

generally, this argument shows that $f[C]$ contains, and therefore equals its own closure. Thus $f[C]$ is closed, and therefore, so is f as a mapping. □

A cautionary remark may be helpful at this point. Historically, point set topology grew out of the study of real and complex analysis. The spaces there all tend to be *Hausdorff* in which distinct points have disjoint neighborhoods. Such spaces are admittedly very nice, but to restrict one's notion of useful topological spaces to those which are Hausdorff is to drastically restrict the richness of the subject. To this day a pro-Hausdorff prejudice can be found to persist amongst some mathematicians. We invite readers to set aside such attitudes (if they have them) when approaching this work. Virtually none of our topological spaces will be Hausdorff.

Chapter 2 which follows next develops needed concepts from model theory. The reader unfamiliar with model theory who wants to linger a bit further with pure topology can first skip ahead to a reading of Chapter 5. In any case, a good general reference for point set topology is Kelley's [15].

Model Theory

2.1. a traditional view

Model theory[1] is one of the few fields of mathematics that attempts a horizontal sweep through all fields of mathematics seeking some global perspective as a result. The only other example of this that comes to my mind is category theory. As a separate discipline model theory found its voice in the late 1940's and early 1950's with the work of Henkin, Robinson and Tarski.

It is typical for a mathematical field to focus on a chosen structure that can be imposed on a set. Typical components of the structure may be special constants or elements of the set, various operations on the set's elements and/or various relations between them. To model the field's object of study, the constants, operations, and relations are required to satisfy various axioms. Specialists in the field then try to classify such models and discover their general properties.

This model theory generalizes. Although not foremost in the minds of specialists, there is usually a sort of trinity involved in their activities. As objects of study, say, the specialist in ordered group theory would ordinarily introduce ordered 4-tuples $\mathbf{G} = (G, <, \cdot, e)$ where G is a nonempty set, $<$ is a binary relation on its elements, $\cdot$ is a binary operation $G \times G \to G$ and $e \in G$ is a distinguished element. The model theorist would term such a 4-tuple $\mathbf{G}$ a *structure*.

To talk about whether or not $\mathbf{G}$ is an ordered group, a *language* $\mathbf{L}$ is needed to discuss $\mathbf{G}$'s properties. For the specialist this is always done informally, but in model theory the language $\mathbf{L}$ is treated as a formal mathematical object on a par with the structure it talks about. As such, $\mathbf{L}$ is a strictly syntactical object. It can be thought of as consisting of an adequate alphabet of formal symbols from which meaningful words can be formed. To begin with there are the "logical symbols" common to all the languages to be considered:

parentheses	), (
variables	$v_0, v_1, \ldots, v_n, \ldots$
connectives	$\vee$ (or) & (and) $\neg$ (not) $\to$ (if then)

---

[1] A good general reference for model theory is [**5**] which the reader can use to suppliment the discussion in this chapter.

quantifiers   ∃ (there exists) ∀ (for all)
special binary   = (equals).

Along with these are the "nonlogical symbols", namely a special formal symbol
for each relation, operator or special individual which forms part of the structure
at hand. In the case of **G** (as throughout this work) we shall cheerfully ignore all
concerns for "use and mention" issues and go straight for whatever notational
abuse seems convenient: Here we shall identify name and thing named, writing
$<$, $\cdot$ and e for both their role as components of **G** and for the symbols in **L**
which name them. In all cases, the context shall uniquely rule as to how such
symbols are to be read.

From **L**'s logical and nonlogical alphabet meaningful types of words or expressions can be built up. First there are *terms* or expressions meant to name
potential elements of **G**'s universe ($= G$). Typical examples are

$$v_{13}, \ e, \ (e \cdot e), \ (v_7 \cdot e),$$
$$((e \cdot (v_{11} \cdot (v_0 \cdot e))) \cdot v_{87}), \ldots$$

An inductive definition of such terms is easily given. Next are the simplest
potential statements one can make about elements of **G**'s universe. These are
called *atomic formulas* and consist of expressions $s = t$ or $s < t$, where s and t
are any terms. From these are built up all other potential statements. Called
formulas, they consist of

  all atomic formulas
  any $(\varphi \vee \psi)$, $(\varphi \ \& \ \psi)$, $(\neg \varphi)$ and $(\varphi \rightarrow \psi)$, where φ, ψ are formulas
  any $(\exists x)\varphi$, $(\forall x)\varphi$, where x is a variable and φ a formula.

An occurrence of a variable x in a formula ψ is *bound* if it is within a subformula
having one of these latter two forms $(\exists x)\varphi$ or $(\forall x)\varphi$. An unbound occurrence of
x in ψ is *free*. A *sentence* of the language **L** is a formula ψ with no free variables
(variables with free occurrence). Similarly, a term t with no variables occurring
in it is said to be *closed*.

So far we have two parts of the trinity alluded to above, namely the *semantic*
part consisting of structures like **G** and the *syntactic* part consisting of formal
languages like **L** corresponding to these structures. The link between the two is
in the notion of *satisfaction*. This will allow us to answer the question of whether
the given structure **G** actually *is* an ordered group or not. The theory of ordered
groups is based on a list of axioms (in this case finite) which must be satisfied.
These can be expressed as formal sentences in the language **L**. For example the
sentences

$$(\forall x)(\forall y)(\forall z)[(x \cdot (y \cdot z) = (x \cdot y) \cdot z]$$
$$(\forall x)(\forall y)[x < y \vee y < x \vee x = y]$$

assert the operator $\cdot$ to be associative and the relation $<$ to satisfy trichotomy.
(Note the multi-layered but cheerful notational abuse just committed. This will
occur repeatedly from now on mostly without comment.)

We can inductively define a relation of satisfaction $\mathbf{G} \models \psi$ between the structure $\mathbf{G}$ and any sentence ψ drawn from its corresponding language $\mathbf{L}$. First we need to expand $\mathbf{L}$ to a language $\mathbf{L}(G)$ which has added individual constant symbols in enough supply to name every element of $\mathbf{G}$'s universe G. We can (with our usual abuse) do this by letting each element $a \in G$ occur in $\mathbf{L}(G)$ as a constant symbol to name itself. At this point every closed term of $\mathbf{L}(G)$ names a unique element in G:

> e names e ; $a \in G$ names itself ;
>
> $s \cdot t$ names the $\cdot$-product of the elements in G
> which are named by s and t, respectively.

If $\psi(x)$ is a formula with at most the variable x free in it, and s is any closed term of $\mathbf{L}(G)$, then $\psi(s)$ shall be the *sentence* obtained by substitution of the symbol s in at each free occurrence of x in $\psi(x)$.

The satisfaction relation is now inductively defined as follows:

$$
\begin{array}{ll}
\mathbf{G} \models s = t & \text{iff the designees of } s \text{ and } t \text{ in } G \text{ are equal} \\
\mathbf{G} \models s < t & \text{iff the designees of } s \text{ and } t \text{ in } G \\
& \quad \text{bear the } < \text{-relation to each other} \\
\mathbf{G} \models \varphi \vee \psi & \text{iff } \mathbf{G} \models \varphi \text{ or } \mathbf{G} \models \psi \\
\mathbf{G} \models \varphi \,\&\, \psi & \text{iff } \mathbf{G} \models \varphi \text{ and } \mathbf{G} \models \psi \\
\mathbf{G} \models \neg\varphi & \text{iff not } \mathbf{G} \models \varphi \\
\mathbf{G} \models \varphi \to \psi & \text{iff either not } \mathbf{G} \models \varphi \text{ or else } \mathbf{G} \models \psi \\
\mathbf{G} \models (\exists x)\varphi(x) & \text{iff for some } a \in G,\, \mathbf{G} \models \varphi(a) \\
\mathbf{G} \models (\forall x)\varphi(x) & \text{iff for all } a \in G,\, \mathbf{G} \models \varphi(a).
\end{array}
$$

With this concept we can now state that the structure $\mathbf{G}$ is an ordered group if $\mathbf{G} \models \psi$ for all axioms ψ of this theory. The *theory* of this subject can be identified with the set T of all sentences in $\mathbf{L}$ which are either axioms or are logical consequences (theorems) of the axioms. If it turned out that the structure $\mathbf{G}$ is an ordered group, then the model theorist would say that $\mathbf{G}$ is a *model* of the theory T of ordered groups.

Model theory studies and gives general constructions for structures $\mathbf{A} = (A, \ldots)$ which are models of theories T expressible in the language $\mathbf{L}$ to which $\mathbf{A}$ corresponds. The reader can easily write down an explicit definition for structures $\mathbf{A}$ in this general sense. Formal languages $\mathbf{L}$ can be similarly defined as well as their correspondence to structures $\mathbf{A}$. The model theorist will speak of $\mathbf{A}$ as a structure *for* $\mathbf{L}$ or as $\mathbf{L}$ being the language *of* $\mathbf{A}$. Structures $\mathbf{A}$ and $\mathbf{A}'$ for the same language $\mathbf{L}$ are said to be of the *same type*.

The structures one can impose on a set A are many. One can keep adding sets, relations, distinguished individuals, and finitary operations, but eventually all possibilities are exhausted. The result is a structure $\mathbf{A} = (A, \ldots)$ we shall call *full*. It contains all elements of A as distinguished individuals, all possible $R \subseteq A^n$ as n-ary relations and all available $f : A^k \to A$ as k-ary operations. If one's mathematical universe of discourse is restricted to the set A, one can think of the full structure $\mathbf{A}$ on A as representing the entirety of mathematical entities there are to discuss. In "non"standandard mathematics such full structures $\mathbf{A}$ play an important role.

Model theory contributes certain highly interesting concepts which give perspective on mathematics in general. These seem not to have arisen naturally in any one specialty yet they apply to virtually all. I will now give brief dissussion of these (and other) notions which arise properly out of model theory.

First, a model theorist will speak of a *monomorphism* between two structures $\mathbf{A} = (A, \dots)$ and $\mathbf{A}' = (A', \dots)$ of the same type with, say, language $\mathbf{L}$. This shall be any mapping $f : A \to A'$ which exactly preserves the truth of atomic sentences in $\mathbf{L}$. Specifically, let $\varphi(x_1, \dots, x_n)$ be any atomic formula of $\mathbf{L}$ whose free variables are among the list of distinct variables $x_1, \dots, x_n$. Let $\mathbf{L}(A)$ and $\mathbf{L}(A')$ be the language enlargements of $\mathbf{L}$ to permit sufficient names for all elements in A or A', respectively. One can inductively define for each closed term t of $\mathbf{L}(A)$ the corresponding closed term $f(t)$ of $\mathbf{L}(A')$ which designates the f-image in A' of the designee of t in A. For f to be a monomorphism $\mathbf{A} \to \mathbf{A}'$ we require for every such atomic formula $\varphi(x_1, \dots, x_n)$ and every choice $t_1, \dots, t_n$ of closed terms in $\mathbf{L}(A)$ that the equivalence[2]

$$\mathbf{A} \models \varphi(t_1, \dots, t_n) \Leftrightarrow \mathbf{A}' \models \varphi(f(t_1), \dots, f(t_n))$$

hold. Here $\varphi(t_1, \dots, t_n)$ is meant to denote the sentence in $\mathbf{L}(A)$ obtained from $\varphi(x_1, \dots, x_n)$ by appropriate substitutions of t's for free x's, and similarly for $\varphi(f(t_1), \dots, f(t_n))$ as a sentence in $\mathbf{L}(A')$. Notice, since $s = t$ is an atomic formula, that any monomorphism f must be a one to one map.

If in the same sense the monomorphism f preserves the exact truth of *all* sentences from $\mathbf{L}(A)$, then f is said to be an *elementary embedding* of $\mathbf{A}$ in $\mathbf{A}'$. If A is a subset of A' and the inclusion map is a monomorphism, then $\mathbf{A}'$ is an *extension* of $\mathbf{A}$ and $\mathbf{A}$ is a *substructure* of $\mathbf{A}'$[3]. When the inclusion is an elementary embedding, one also speaks of *elementary extensions* and *elementary substructures*. Two structures $\mathbf{A}$ and $\mathbf{A}'$ of the same type are said to be *elementarily equivalent* if they satisfy the same sentences from their common language $\mathbf{L}$.

Given a structure $\mathbf{A} = (A, \dots)$ we mean by its *power* the cardinality of its universe A. We shall also speak of the *cardinality* of a language $\mathbf{L}$ and mean the cardinality of the set of its formulas. The structure $\mathbf{A} = (A, \dots)$ is *universal of power* κ if every structure $\mathbf{A}' = (A', \dots)$ of power less than κ which is elementarily equivalent to $\mathbf{A} = (A, \dots)$ has an elementary embedding into $\mathbf{A}$.

Given a structure $\mathbf{A} = (A, \dots)$ and an index set I, we consider the possible *I-types compatible with* $\mathbf{A}$. We let T be the theory consisting of all sentences

---

[2]If only the left to right implication holds one refers to f as *homomorphism* from $\mathbf{A}$ to $\mathbf{A}'$.

[3]If the inclusion map of A into A' is only a homomorphism of $\mathbf{A}$ into $\mathbf{A}'$, one calls $\mathbf{A}'$ a *homomorphic extension* of $\mathbf{A}$.

from **L** which are satisfied by **A**. Expand **L** to a further language **L**(I) so that each element $i \in I$ now occurs in **L**(I) as a new individual constant symbol. An I-type compatible with **A** shall be a maximal collection t of sentences in **L**(I) which contains T, is consistent (contains no contradictions φ & $\neg\varphi$), and yet is deductively closed (any sentence provable from sentences in t also lies in t). An element f of the cartesian product A^I *realizes* the type t if for every sentence $\varphi \in t$ one has **A** $\models \varphi'$, where φ' is the sentence, in a suitably enlarged language **L**(A), which is obtained from φ by replacing each symbol $i \in I$ occurring in φ by its f-related constant symbol $f(a)$ in **L**(A). When this happens t is said to be the type *of f* and is written $t(f)$. The "arity" of the type t is the cardinality κ of the index set I.

The structure **A** $= (A, \dots)$ is said to be κ-*homogeneous* if for any index set I of cardinality less than κ and any cartesian elements $f, g \in A^I$ which realize the same I-type, the possible types realized by unary extensions of either f or g are the same. Put otherwise, if $I' = I \cup \{*\}$ where $*$ is new and if $f' \in A^{I'}$ is an arbitrary extension of f, then g has an extension $g' \in A^{I'}$ realizing the same type as f' (and conversely for extensions of g).

We come finally to the model theorist's notion of saturation. Let **A** $= (A, \dots)$ be a structure and κ a cardinal number. The structure **A** is said to be κ-*saturated* if for every subset $D \subseteq A$ of cardinality less than κ, if we let **A**(D) be the expansion of the structure **A** so that each element $d \in D$ now becomes a distinguished individual listed in the new structure, then each 1-ary type compatible with **A**(D) is realized by an element $a \in A$.

It is a standard result in model theory that any structure **A** $= (A, \dots)$ can be elementarily embedded in a structure **A**$' = (A', \dots)$ whose level of saturation can be chosen to be arbitrarily high. The reader may consult [**5**] for a proof of this fact. It forms a nice example of a result from model theory which is valid (and useful) in a broad spectrum of mathematical fields, yet apparently was never specifically proved by a specialist in any one of them. The application to the case when the original structure **A** $= (A, \dots)$ is *full* is of particular importance in "non"standard mathematics. In a sense, the extending structure **A**$' = (A', \dots)$ enlarges one's mathematical universe in a manner in which mythological mathematical objects (e.g., infinite prime numbers) find realization and can be taken seriously.

2.2. topological connections

The above discussion is more or less how a model theorist would phrase matters. Ironically, the visitor finds that much of this reduces to point set topology. As Marshall Stone is reported to have once said: "One must always topologize." I will do just that. Indeed the discussion which follows will serve as an introduction to the fundamental link I shall be making later between topology and "non"standard mathematics.

Let $\mathbf{A} = (A, \dots)$ be a structure with language $\mathbf{L}$. We give each finitary power A^n a zero dimensional topology by taking as an open basis all subsets $R \subseteq A^n$ which are $\mathbf{A}$-definable. By this we mean there exists a formula $\varphi(x_1, \dots, x_n)$ of $\mathbf{L}$ whose free variables are among the distinct variables $x_1, \dots, x_n$, and for which

$$R = \{(a_1, \dots a_n) \in A^n : \mathbf{A} \models \varphi(a_1, \dots, a_n)\}.$$

Such definable subsets are closed under finite unions, finite intersections, under complements, and cartesian products. Furthermore, if $\pi : A^n \to A^m$ is any projection or permutation (with $n = m$), then these sets are closed under direct π-images as well. Clearly the topologies thus imposed on the A^n are all zero dimensional, are "symmetrical" in the sense discussed in Chapter 1, and all projection maps $\pi : A^n \to A^m$ are continuous. As described in Chapter 1, all powers A^I now receive canonical topologies, which we will call the $\mathbf{A}$-*topologies*. A subset $R \subseteq A^I$ which is the inverse image by a "projection" $\pi : A^I \to A^n$ of some $\mathbf{A}$-definable set in A^n will be called an $\mathbf{A}$-*basis set*.

Let us reconsider the notion of I-types compatible with $\mathbf{A}$. Such an I-type t is essentially nothing more than a maximal family of $\mathbf{A}$-basis sets in A^I satisfying the finite intersection property. Each element of t can be written as $\varphi(i_1, \dots, i_n)$ for some $i_1, \dots, i_n \in I$ and some formula $\varphi(x_1, \dots, x_n)$ of $\mathbf{L}$ whose free variables are among the distinct variables $x_1, \dots, x_n$. To this sentence we let correspond the $\mathbf{A}$-basis set

$$\{f \in A^I : \mathbf{A} \models \varphi(f(i_1), \dots, f(i_n))\}.$$

The correspondence of t to such a family of $\mathbf{A}$-basis sets clearly characterizes t exactly. From this it follows for any $f, g \in A^I$ that f and g realize the same I-type iff $\overline{\{f\}} = \overline{\{g\}}$ in the $\mathbf{A}$-topologies.

The following three propositions further illustrate the relevance of this topological perspective:

PROPOSITION 2.1. *Let* $\mathbf{A} = (A, \dots)$ *be any structure and* κ *an infinite cardinal number. If* A^I *is* $\mathbf{A}$-*compact for every set* I *of cardinality less than* κ, *then* $\mathbf{A}$ *is* κ-*universal. Conversely, if* A *is* κ-*universal and the cardinality of the language* $\mathbf{L}$ *of* $\mathbf{A}$ *is less than* κ, *then for every set* I *of cardinality less than* κ, A^I *is* $\mathbf{A}$-*compact* [4].

PROOF. Assume A^I is $\mathbf{A}$-compact for every set I of cardinality less than κ and let $\mathbf{A}' = (A', \dots)$ be a structure of cardinality less than κ which is elementarily equivalent to $\mathbf{A}$. Let $I = A'$ and let t be the I-type of all sentences φ in $\mathbf{L}(I) = \mathbf{L}(A')$ for which $\mathbf{A}' \models \varphi$ is the case. Since $A^I = A^{A'}$ is $\mathbf{A}$-compact, this type is realized by some $f \in A^I$ which is, clearly, a map $f : A' \to A$ giving an elementary embedding of $\mathbf{A}'$ into $\mathbf{A}$.

---

[4]A full equivalence of compactness and universality is impossible, given the traditional definition of the latter. Indeed *any* infinite structure $\mathbf{A}$, having no finite elementary equivalents, is, by default, ω-universal ($\omega =$ the first infinite cardinal).

The converse portion of this proposition is well-known. It can be found in [**5**, exercise 4.3.24] expressed in traditional model theoretic langauge. □

PROPOSITION 2.2. *Let* $\mathbf{A} = (A, \dots)$ *be any structure and* κ *an infinite cardinal number. Then* $\mathbf{A}$ *is* κ-*homogeneous iff for every subset* $I \subseteq \kappa$ *of cardinality below* κ *the projection map* $\pi : A^\kappa \to A^I$ *preserves* $\mathbf{A}$-*closures of points.*

PROOF. Assume that each of the named projection maps preserve $\mathbf{A}$-closures of points. We then show that $\mathbf{A}$ is κ-homogeneous. Let $I \subseteq I'$ be any sets of cardinality less than κ and let elements $f, g \in A^I$ realize the same I-type, that is, $\overline{\{f\}} = \overline{\{g\}}$ in the $\mathbf{A}$-topologies. Let f' be any extension of f to an element of $A^{I'}$. For κ-homogeneity we need to find an extension g' of g to an element of $A^{I'}$ which realizes the same I'-type as f'. We can of course assume that I and I' are both subsets of κ. Since the composition of the projections $A^\kappa \to A^{I'}$ and $A^{I'} \to A^I$ results in the projection $A^\kappa \to A^I$ and since all three are onto maps and the first and third preserve closures of points, it follows that the projection $\pi' : A^{I'} \to A^I$ has the same property. Thus $\pi'[\overline{\{f'\}}] = \overline{\{f\}}$. But $g \in \overline{\{f\}}$ and hence, there exists $g' \in \overline{\{f'\}}$ such that $\pi'[g'] = g$. Clearly, g' realizes the same type as f', and $\mathbf{A}$ is seen to be κ-homogeneous.

Now suppose, conversely, that $\mathbf{A}$ is κ-homogeneous. Let $I \subseteq \kappa$ be of cardinality smaller than κ and let $f, g \in A^I$ satisfy $\overline{\{f\}} = \overline{\{g\}}$. To show $\pi : A^\kappa \to A^I$ preserves closures of points we need to assume some $f' \in A^\kappa$ extends f and show how to find a $g' \in A^\kappa$ extending g for which $g' \in \overline{\{f'\}}$. We can write $\kappa = \bigcup_{\alpha < \kappa} I_\alpha$ where $I_0 = I$, and for $\alpha < \kappa$, $I_{\alpha+1}$ is I_α plus a single new element, and $I_\alpha = \bigcup_{\alpha' < \alpha} I_{\alpha'}$ if α is a limit ordinal. Clearly, for each $\alpha < \kappa$ the cardinality of I_α is less than κ. By induction, we form for each $\alpha \leq \kappa$ an element g_α in the $\mathbf{A}$-closure of $\{f'|_{I_\alpha}\}$ such that its restriction to $I = I_0$ is g. When $\alpha = \alpha' + 1$ $(= \alpha' \cup \{\alpha'\})$ is a successor ordinal (and hence $\alpha < \kappa$) we use $\mathbf{A}$'s κ-homogeneity to construct g_α from $g_{\alpha'}$. For limit α, we let g_α be the unique common extension of the $g_{\alpha'}$ for $\alpha' < \alpha$ and use Proposition 1.1 to confirm correctness of this choice. Clearly, the final g' we want is g_κ. □

PROPOSITION 2.3. *Let* κ *be an infinite cardinal. Then a structure* $\mathbf{A} = (A, \dots)$ *is* κ-*saturated iff* A^κ *is* $\mathbf{A}$-*compact and all projections* $\pi : A^\kappa \to A^I$ *with* $I \subseteq \kappa$ *of cardinality less than* κ *preserve closures of points. When this is the case, each of these projections, in fact, is a closed mapping.*

PROOF. For κ greater than the power of the language $\mathbf{L}$ of $\mathbf{A}$, this is just a topological restatement of the well-known characterization of saturation in terms of universality and homogeneity (see [**5**, page 299]). In any case, the fact that the projections will be closed mappings follows immediately from Proposition 1.2.

To prove the claim for arbitrary κ, we note that there is an easily seen equivalent formulation of κ-saturation for $\mathbf{A}$, namely, that for any sets I, $I' = I \cup \{*\}$ with I of cardinality less than κ, that if $\pi' : A^{I'} \to A^I$ is the obvious projection map, then for any $f \in A^I$, $\pi^{-1}(f) \subseteq A^{I'}$ is $\mathbf{A}$-compact.

Suppose now that A^κ is **A**-compact and **A** is κ-homogeneous. Let I, $I' = I \cup \{*\}$ be as described. Suppose also that $f \in A^I$ and $\{U_i\}_{i \in \Gamma}$ is a family of **A**-basis clopens in $A^{I'}$ such that

$$\{U_i \cap \pi^{-1}(f)\}_{i \in \Gamma}$$

satisfies the finite intersection property. Then the family

$$\{U_i \cap \pi^{-1}[\overline{\{f\}}]\}_{i \in \Gamma}$$

also satisfied the finite intersection property, and consists of closed sets in $A^{I'}$. Since A^κ is **A**-compact, so is $A^{I'}$ and hence,

$$f' \in \bigcap_{i \in \Gamma} U_i$$

must exist so that $\pi(f') \in \overline{\{f\}}$. Since the intersection of the U_i's is closed, we have

$$\overline{\{f'\}} \subseteq \bigcap_{i \in \Gamma} U_i,$$

and, by κ-homogeneity, f' can be chosen so that $f' \in \pi^{-1}(f)$. Thus, $\pi^{-1}(f)$ is **A**-compact and more generally, **A** is κ-saturated.

Suppose, conversely, that **A** is κ-saturated. Then by [**5**, page 297] **A** is κ-homogeneous. From Proposition 2.1 we "almost" know that A^κ is **A**-compact, since by [**5**, page 298] **A** is also κ^+-universal. In any case, suppose $\{U_i\}_{i \in \Gamma}$ is a family of **A**-basis clopens $U_i \subseteq A^\kappa$ satisfying the finite intersection property. For each $\alpha \leq \kappa$, let $\pi_\alpha : A^\kappa \to A^\alpha$ be the obvious projection. Then for each α, $\{U_i^\alpha\}_{i \in \Gamma}$, where $U_i^\alpha = \pi_\alpha[U_i]$, is an f.i.p. family of sets in A^α. Since finitary projections preserve **A**-basis clopens, it follows easily that π does as well, so in fact, the U_i^α are *closed*. As in the proof of Proposition 2.2, we can inductively form a mutually extending sequence of functions g_α, for $\alpha \leq \kappa$, such that in every case

$$g_\alpha \in \bigcap_{i \in \Gamma} U_i^\alpha .$$

For $\alpha = 0$ we use the fact, clearly implied by κ-saturation, that A is **A**-compact. We use the above equivalent formulation of κ-saturation to derive $g_{\alpha+1}$ from g_α (i.e., set I, $I' = \alpha$, $\alpha + 1$ with $* = \{\alpha\}$). At limit ordinals α, g_α may be taken as the common extension of all the $g_{\alpha'}$'s, for $\alpha' < \alpha$. Since $g_\kappa \in A^\kappa$ belongs to all of the U_i's, this shows that A^κ is **A**-compact. $\square$

2.3. a generalization

The foregoing has been an exploration of traditional model theory. For the purposes of applications to "non"standard mathematics, a looser less finely tuned theory — something I shall term *point set model theory* — will be more appropriate. In this theory, syntactical considerations (languages) and semantical issues (satisfaction of sentences) will be nominally suppressed, although they will be readily called forth as needed. We make the structure **A** as the basic object and in fact loosen the notion of what a structure shall mean. The structure **A** shall now consist of its domain A and what we have been calling the **A**-basis sets in the finitary powers of A. Specifically, $\mathbf{A} = (A, \underline{A})$ where A is a nonempty set and $\underline{A} = (\underline{A}_1, \underline{A}_2, \underline{A}_3, \dots)$ where for each natural number n, $\underline{A}_n \subseteq \mathcal{P}(A^n)$ (= power set of A^n) is a nonempty collection of n-ary relations on A which is closed under finite unions, finite intersections and *relative* complements. We also require that

> A be the union of elements in $\underline{A}_1$
> for each $A_0 \in \underline{A}_1$ the *diagonal* set $\Delta_{A_0} = \{(a,a) : a \in A_0\}$
>   is an element of $\underline{A}_2$
> the cartesian products of elements from $\underline{A}_n$ and $\underline{A}_m$
>   lie in $\underline{A}_{n+m}$
> for every finitary projection $\pi : A^n \to A^m$, the direct
> π-images of elements in $\underline{A}_n$ lie in $\underline{A}_m$
> for every finitary permutation $\sigma : A^n \to A^n$, the direct
> σ-images of elements in $\underline{A}_n$ lie in $\underline{A}_n$.

When we need to distinguish such a structure **A** from the model theorist's traditional version we will call **A** a *point set* structure. As before, sets or relations from the collections $\underline{A}_n$ will be called **A**-*basis* sets or relations.

Notice that the domain A is not itself assumed to be an **A**-basis set. When this is the case we refer to **A** as a *local* (point set) structure. Otherwise we say that **A** is *non*local. The reason for this distinction is that in several versions of "non"standard mathematics (e.g., Robinson's enlargements and the "non"standard set theories of Hrbáček), the model theory required is *non*local. In these contexts, as we shall see below, quantification or saturation assumptions are only applied locally.

It is important for us that these generalized point set structures be closed under direct limits. For this we need morphisms. A *morphism* from a structure $\mathbf{A} = (A, \underline{A})$ to structure $\mathbf{A}' = (A', \underline{A}')$ shall consist of a pair $\mathbf{f} = (f, \underline{f})$, where $f : A \to A'$ is an ordinary mapping of the sets, and $\underline{f} = (\underline{f}_1, \underline{f}_2, \underline{f}_3, \dots)$ consists, for each n, of a map $\underline{f}_n : \underline{A}_n \to \underline{A}'_n$, so that, if we let $f_n : A^n \to A'^n$ be the map sending $(a_1, a_2, \dots, a_n)$ to $(f(a_1), f(a_2), \dots, f(a_n))$, then the following properties are satisfied:

> $f_n[R] \subseteq \underline{f}_n(R)$ for all $R \in \underline{A}_n$
> $\underline{f}_n$ preserves unions, intersections and relative complements
> $\underline{f}_2(\Delta_{A_0}) = \Delta_{\underline{f}_1(A_0)}$ for each $A_0 \in \underline{A}_1$
> the $\underline{f}_n$'s collectively preserve cartesian products and direct
>   images by finitary projection and permutation maps.

Composition of these morphisms is defined in the obvious manner.

A *directed set* shall be any pair $(\Gamma, <)$ where Γ is a nonempty set and $<$ a partial ordering of Γ for which, to each $i, j \in \Gamma$, there exists $k \in \Gamma$ such that

$i, j < k$. A *directed system* of point set structures and their morphisms *indexed* by $(\Gamma, <)$ is a collection $\{\mathbf{A}(i) : i \in \Gamma\}$ of point set structures and a collection $\{\mathbf{f}_{ij} : i, j \in \Gamma, i < j\}$ of their morphisms where, for $i, j \in \Gamma$ with $i < j$, $\mathbf{f}_{ij} = (f_{ij}, \underline{f}_{ij})$ is a morphism $\mathbf{A}(i) \to \mathbf{A}(j)$. These morphisms are also to satisfy a compatibility: for $i, j, k \in \Gamma$ with $i < j < k$ the composition $\mathbf{f}_{jk} \circ \mathbf{f}_{ij}$ is to coincide with $\mathbf{f}_{ik}$.

We construct the *direct limit* $\mathbf{A} = (A, \underline{A})$ of these $\mathbf{A}(i) = (A(i), \underline{A}(i))$ for $i \in \Gamma$ as follows. Without loss of generality, we can assume in what follows that the $A(i)$'s are all disjoint. Identify $a \in A(i)$ with $b \in A(j)$ if there exists $i, j < k \in \Gamma$ such that $f_{ik}(a) = f_{jk}(b)$. Let A be the union of all the $A(i)$'s modulo such identifications, and let $f(i) : A(i) \to A$ be the canonical map sending $a \in A(i)$ to its equivalence class in A. Clearly, for $i < k \in \Gamma$ we have $f(k) \circ f_{ik} = f(i)$. To each $R \in \underline{A}_n(i)$, $f_n(i)[R] \subseteq A^n$ is a well defined subset. We define $\underline{f}_n(i)(R)$ to be the union, for all $i < k \in \Gamma$, of the $f_n(k)[(\underline{f}_{ik})_n(R)]$. We let $\underline{A}_n$ consist of all $\underline{f}_n(i)(R) \subseteq A^n$, for $i \in \Gamma$ and $R \in \underline{A}_n(i)$. It is routine to check that $\mathbf{A} = (A, \underline{A})$ with $\underline{A} = (\underline{A}_1, \underline{A}_2, \dots)$ is a point set structure, that for each $i \in \Gamma$, $\mathbf{f}(i) = (f(i), \underline{f}(i))$, with $\underline{f}(i) = (\underline{f}_1(i), \underline{f}_2(i), \dots)$, is a morphism $\mathbf{A}(i) \to \mathbf{A}$, and furthermore that, for all $i < k \in \Gamma$, $\mathbf{f}(k) \circ \mathbf{f}_{ik} = \mathbf{f}(i)$.

The structure $\mathbf{A} = (A, \underline{A})$ is called a direct limit since it satisfies the following *universal mapping principle* (a standard item from category theory):

> For any system of morphisms $\mathbf{g}(i) : \mathbf{A}(i) \to \mathbf{B}$ for which $\mathbf{g}(i) = \mathbf{g}(k) \circ \mathbf{f}_{ik}$ for all $i < k \in \Gamma$, there exists a unique morphism $\mathbf{g} : \mathbf{A} \to \mathbf{B}$ such that $\mathbf{g}(i) = \mathbf{g} \circ \mathbf{f}(i)$ for each $i \in \Gamma$.

If our point set structures generalize those of conventional model theory, then it is also true that our notion of morphisms generalize the idea of elementary embeddings. Borrowing language developed in "non"standard analysis we will express this phenomenon by saying that morphisms of point set structures satisfy a (local) *transfer principle*.

Indeed, let $\mathbf{A} = (A, \underline{A})$ be an arbitrary point set structure. An element $a \in A$ will be called an *individual* of $\mathbf{A}$ if its singleton $\{a\}$ is an $\mathbf{A}$-basis set (element of $\underline{A}_1$). A *local function* of $\mathbf{A}$ will be any finitary function $h : A^n \to A^m$ where for each $A' \in \underline{A}_n$ the graph $G(h|_{A'})$ of the restriction of h to A' is an element of $\underline{A}_{n+m}$. For each $n, m > 1$, we let $F_{n,m}(\mathbf{A})$ be the collection of all local functions $A^n \to A^m$ belonging to $\mathbf{A}$. The function families $F_{n,m}(\mathbf{A})$ have the following obvious properties:

> all projection maps and identity maps are included in
>   the appropriate $F_{n,m}(\mathbf{A})$'s
> for all $f \in F_{n,m}(\mathbf{A})$ and $g \in F_{m,k}(\mathbf{A})$, $g \circ f \in F_{n,k}(\mathbf{A})$
> for all $f \in F_{n,m}(\mathbf{A})$ and $g \in F_{n,k}(\mathbf{A})$, $< f, g > \in F_{n,m+k}(\mathbf{A})$
> for all $f \in F_{n,m}(\mathbf{A})$ and $g \in F_{k,l}(\mathbf{A})$, $f \times g \in F_{n+k,m+l}(\mathbf{A})$.

Here $g \circ f$ represents the composition of f with g and $< f, g >$, and $f \times g$ are defined by the equations

$$< f, g > (a) \ = \ < f(a), g(a) >$$
$$(f \times g)(a, b) \ = \ < f(a), g(b) > .$$

To the point set structure $\mathbf{A}$ we assign the language $\mathbf{L}$ whose individual constant symbols, n-ary function symbols and n-ary relation symbols, respectively, are the individuals of $\mathbf{A}$, the local n-ary functions $h \in F_{n,1}(\mathbf{A})$ and n-ary $\mathbf{A}$-basis relations $R \in \underline{A}_n$. A formula φ of $\mathbf{L}$ is *local* if all of its quantifiers are of the form $(\exists x \in A_0)$ or $(\forall x \in A_0)$ with $A_0 \in \underline{A}_1$ an $\mathbf{A}$-basis set. Such quantifiers are abbreviations where

$$(\exists x \in A_0)\psi \text{ means } (\exists x)[A_0(x) \ \& \ \psi]$$
$$(\forall x \in A_0)\psi \text{ means } (\forall x)[A_0(x) \rightarrow \psi].$$

Now let $\mathbf{A}' = (A', \underline{A}')$ be a second point set structure and $\mathbf{f} = (f, \underline{f})$ a morphism $\mathbf{A} \rightarrow \mathbf{A}'$. Let $\mathbf{A}'$ have language $\mathbf{L}'$. It is clear for any individual constant symbol a in $\mathbf{L}$ that

$$\{a\} \times \{a\} = \Delta_{\{a\}}, \text{hence}$$
$$\underline{f}_1(\{a\}) \times \underline{f}_1(\{a\}) = \underline{f}_2(\{a\} \times \{a\}) = \Delta_{\underline{f}_1(\{a\})}$$

and since $\{f(a)\} \subseteq \underline{f}_1(\{a\})$, that $\{f(a)\} = \underline{f}_1(\{a\})$ is forced, so that $a' = f(a)$ is also an individual constant symbol for $\mathbf{L}'$. Letting n-ary relation symbol $R \in \underline{A}_n$ in $\mathbf{L}$ correspond to the n-ary relation symbol $R' = \underline{f}_n(R) \in \underline{A}'_n$ in $\mathbf{L}'$ we get, more generally, a *partial interpretation* of the first language in the second. It turns out that if the morphism $\mathbf{f} : \mathbf{A} \rightarrow \mathbf{A}'$ of point set structures is *dominant*, that is, if

to every $A'_0 \in \underline{A}'_1$ there exists $A_0 \in \underline{A}_1$
such that $A'_0 \subseteq \underline{f}_1(A_0)$,

then for each function $h \in F_{n,m}(\mathbf{A})$ there exists a unique function $h' \in F_{n,m}(\mathbf{A}')$ such that $\underline{f}_{n+m}(A_0) \subseteq G(h')$ whenever $A_0 \in \underline{A}_{n+m}$ satisfies $A_0 \subseteq G(h)$. When this is the case we get, by letting h' correspond to h, a *full* interpretation of the language $\mathbf{L}$ of $\mathbf{A}$ in the language $\mathbf{L}'$ of $\mathbf{A}'$.

The importance of this is the following:

THEOREM 2.1. [the transfer principle] *Let* $\mathbf{f} : \mathbf{A} \rightarrow \mathbf{A}'$ *be a morphism of point set structures with languages* $\mathbf{L}$ *and* $\mathbf{L}'$ *as above. Let* $\varphi(\mathbf{x})$ *be a local formula of* $\mathbf{L}$ *without function symbols whose free variables are among the distinct variables* $\mathbf{x} = x_1, x_2, \ldots, x_n$. *Let* $\varphi'(\mathbf{x})$ *be the* $\mathbf{f}$*-determined interpretation in* $\mathbf{L}'$ *of the formula* $\varphi(\mathbf{x})$. *Then for any elements* $\mathbf{a} = a_1, a_2, \ldots, a_n \in A$, *if we let* $\mathbf{a}' = a'_1, a'_2, \ldots, a'_n$ *with* $a'_i = f(a_i)$ *for* $i = 1, 2, \ldots, n$, *then*

$$\mathbf{A} \models \varphi(\mathbf{a}) \text{ iff } \mathbf{A}' \models \varphi'(\mathbf{a}').$$

If the morphism $\mathbf{f} : \mathbf{A} \rightarrow \mathbf{A}'$ *is dominant then this property holds for arbitrary local formulas* $\varphi(\mathbf{x})$ *in the language* $\mathbf{L}$.

PROOF. One argues by induction on complexity that for each *local* formula $\varphi(\mathbf{x})$ from $\mathbf{L}$ with $\mathbf{x}$ n-ary one has for any $R \in \underline{A}_n$ that

$$\{\mathbf{a} \in R : \mathbf{A} \models \varphi(\mathbf{a})\} \in \underline{A}_n.$$

A similar induction shows that if such local $\varphi(\mathbf{x})$ contains no function symbols then

$$(*) \quad \underline{f}_n(\{\mathbf{a} \in R : \mathbf{A} \models \varphi(\mathbf{a})\}) = \{\mathbf{a}' \in \underline{f}_n(R) : \mathbf{A}' \models \varphi'(\mathbf{a}')\}.$$

The definitions given easily imply that for any $R' \in \underline{A}_n$ one has

$$f_n[R'] = \underline{f}_n(R') \cap f_n[A^n],$$

and from this the transfer theorem for local $\varphi(\mathbf{x})$ without function symbols is immediate. This partial theorem can now be used in the case when $\mathbf{f}$ is dominant to justify the full interpretation of $\mathbf{L}$ in $\mathbf{L}'$. A further induction in this case establishes $(*)$ for *all* formulas from $\mathbf{L}$ and, in a similar manner, the rest of the transfer theorem follows. $\square$

We will say that a structure $\mathbf{A}' = (A', \underline{A}')$ is a *substructure* of a structure $\mathbf{A} = (A, \underline{A})$ if $A' \subseteq A$, and for each n, $\underline{A}'_n \subseteq \underline{A}_n$ holds. Clearly, when this happens the appropriate inclusion maps make up a morphism $\mathbf{A}' \to \mathbf{A}$. The substructure $\mathbf{A}'$ is *full in* $\mathbf{A}$ (alternatively, $\mathbf{A}'$ is a *full substructure* of $\mathbf{A}$) if for all n

$$R' \in \underline{A}'_n, R \in \underline{A}_n \text{ and } R \subseteq R' \text{ imply that } R \in \underline{A}'_n.$$

Obviously, any structure $\mathbf{A}$ is a canonical direct limit of its local full substructures.

If a structure $\mathbf{A}$ is full in every context in which it is a substructure, we say that $\mathbf{A}$ itself is *full* as a structure. This is equivalent to saying for all n that

$$\mathcal{P}(R) \subseteq \underline{A}_n \text{ for each } R \in \underline{A}_n \ (\mathcal{P}(R) = \text{power set of } R).$$

This notion generalizes the full structures of traditional model theory mentioned in Section 2.1 and plays a important role in "non"standard mathematics.

Given a structure $\mathbf{A} = (A, \underline{A})$, we can introduce $\mathbf{A}$-topologies on each of the cartesian powers A^I as done previously. Each of the finitary powers A^n is given the collection $\underline{A}_n$ as an open basis and since the resulting topologies are "symmetric", all the A^I are canonically topologized as well. We will say that the structure $\mathbf{A}$ is κ-*saturated* if each of its local full substructures are κ-saturated in the sense of Proposition 2.3. We may similarly use Propositions 2.1 and 2.2 to define when $\mathbf{A}$ is κ-*universal* or κ-*homogeneous*. It will be useful to include a notion of when A is *locally* κ-saturated, namely, when $\mathbf{A}$ is at least covered by local full substructures which are κ-saturated. This will be relevant later in the Robinson and Hrbáček versions of "non"standard mathematics.

A useful observation is the following:

PROPOSITION 2.4. *For any structure* $\mathbf{A} = (A, \underline{A})$, *its local functions are* $\mathbf{A}$-*continuous.*

PROOF. Let $f : A^n \to A^m$ be an element of $F_{n,m}(\mathbf{A})$, let $R' \in \underline{A}_m$ be any basis open $\subseteq A^m$ and let $q \in f^{-1}[R'] \subseteq A^n$ be arbitrary. For continuity we need to find basis open $q \in R \in \underline{A}_n$ such that $f[R] \subseteq R'$. Picking any basis open $q \in S \in \underline{A}_n$ we can choose R to be $\pi[G(f|_S) \cap S \times R']$ where $\pi : A^{n+m} \to A^n$ is the projection which omits the last m coordinates. $\square$

By contrast, it is quite possible for a morphism $\mathbf{f} : \mathbf{A} \to \mathbf{A}'$ of point set structures to be unrelated to the topologies involved. An important exception is the following. We may always associate to $\mathbf{f}$ its "image" substructure $\mathbf{f}[\mathbf{A}] \subseteq \mathbf{A}'$ where $\mathbf{f}[\mathbf{A}] = (A'', \underline{A}'')$ with

$$A'' = \bigcup\nolimits_{A_0 \in \underline{A}_1} \underline{f}_1(A_0) \subseteq A'$$
$$\underline{A}''_n = \underline{f}_n[\underline{A}_n] \subseteq \underline{A}'_n.$$

We will call a morphism $\mathbf{f} : \mathbf{A} \to \mathbf{A}'$ *proper* if its image $\mathbf{f}[\mathbf{A}]$ is full in $\mathbf{A}'$.

PROPOSITION 2.5. *Assume that* $\mathbf{f} : \mathbf{A} \to \mathbf{A}'$ *is a proper morphism of point set structures. Then*

    a) *for each* n, $f_n : A^n \to A'^n$ *is an immersion of topological spaces*
    b) *for each* n *and* $R \in \underline{A}_n$, $\underline{f}_n(R) = \overline{f_n[R]}$.

PROOF. Clearly, for any $\mathbf{f} : \mathbf{A} \to \mathbf{A}'$ one has for all n and $R \in \underline{A}_n$ that

$$\overline{f_n[R]} \subseteq \underline{f}_n(R).$$

Assume $\mathbf{f}[\mathbf{A}] \subseteq \mathbf{A}'$ is a full substructure and suppose

$$\overline{f_n[R]} \subsetneq \underline{f}_n(R).$$

Since the former is closed and the latter clopen, their difference is nonempty and open. Thus nonempty $S' \in \underline{A}'_n$ exists for which

$$S' \subseteq \underline{f}_n(R) - \overline{f_n[R]}.$$

Since $\mathbf{f}$ is proper we can find $S \in \underline{A}_n$ for which $S' = \underline{f}_n(S)$. By transfer

$$\emptyset \neq S \subseteq R.$$

But then $f_n[S] \subseteq S' \cap f_n[R]$, which contradicts the fact that the latter is empty. This proves (b).

Since for any $R \in \underline{A}_n$ one has

$$f_n[R] = \underline{f}_n(R) \cap f_n[A^n],$$

so the former is a subspace clopen in $f_n[A^n]$. Thus, each f_n is one to one and has a continuous inverse. To show it is an immersion we need only show it is continuous. Let $R' \in \underline{A}'_n$ and choose arbitrary $\mathbf{p} \in f_n^{-1}[R'] \subseteq A^n$. Pick $\mathbf{p} \in S \in \underline{A}_n$. Since $\mathbf{f}$ is proper we can find some $S_0 \in \underline{A}_n$ for which

$$\underline{f}_n(S) \cap R' = \underline{f}_n(S_0).$$

Clearly, $\mathbf{p} \in S_0 \subseteq f_n^{-1}[R']$. Thus, the latter is open and f_n is continuous. $\square$

An easy observation is that a morphism $\mathbf{f} : \mathbf{A} \to \mathbf{A}'$ is both proper and dominant precisely when $\mathbf{f}[\mathbf{A}] = \mathbf{A}'$. For the sake of completeness we also include the following result:

THEOREM 2.2. *To each point set structure* $\mathbf{A}$ *and infinite cardinal* κ *there exists a proper dominant morphism* $\mathbf{A} \to \mathbf{A}'$ *of* $\mathbf{A}$ *into a point set structure which is* κ-*saturated.*

PROOF. Consider $\mathbf{A}$ as a conventional model theoretic structure in the obvious manner. Elementarily extend $\mathbf{A}$ to a κ-saturated conventional model theoretic structure $\mathbf{A}'$. Reinterpret $\mathbf{A}$ and $\mathbf{A} \subseteq \mathbf{A}'$ in terms of point set model theory. $\square$

2.4. internal domains

Clearly, the foregoing generalization of model theory which I have sketched can be explored further, but for our purposes I shall now narrow attention to a special class of point set structures which are relevant to "non"standard mathematics. These structures will be called "internal domains".

For any structure $\mathbf{A} = (A, \underline{A})$ one can ask to what extent the $\mathbf{A}$-topologies determine the structure itself. Although the $\mathbf{A}$-basis relations are clopens in the finitary $\mathbf{A}$-topologies, it would be rare for the converse to hold. We will call a morphism $\mathbf{A} \to \mathbf{A}' = (A', \underline{A}')$ *topologically conservative* if for all n, the induced maps $A^n \to A'^n$ are homeomorphisms. Obviously, in this case the $\mathbf{A}$-topologies and $\mathbf{A}'$-topologies are essentially the same, although as a structure $\mathbf{A}'$ will have a better chance to be determined by these topologies than $\mathbf{A}$ does. If no proper topologically conservative extension of $\mathbf{A}$ exists, we shall say that $\mathbf{A}$ is *topologically maximal.* By the axiom of choice, topologically maximal extensions always exist, but of course they may not be unique. In the case of internal domains, which we define next, matters will be quite different.

An *internal pre-domain* shall be any point set structure $\mathbf{A}$ for which there exists a morphism $* : {}^\circ\mathbf{A} \to \mathbf{A}$ from a full structure ${}^\circ\mathbf{A}$ for which the associated map $*_1 : {}^\circ\underline{A}_1 \to \underline{A}_1$ is a bijection. This notion obviously generalizes the notion in traditional model theory of a structure which elementarily extends a full structure. As remarked above, when combined with saturation, such extensions permit mathematical mythology to have respectful realization.

Essentially, an internal domain is to be an internal pre-domain with the best chance to be topologically self-determined. Specifically, an **internal domain** shall be any internal pre-domain which is topologically maximal in the sense that any topologically conservative morphism $\mathbf{A} \to \mathbf{A}'$ into a second internal pre-domain is necessarily an isomorphism of the point set structures.

At this point we have arrived at a first view of the central concept of this work. To give some initial familiarity with its definition, I shall include here the following preliminary results[5]:

PROPOSITION 2.6. *Any* local *internal pre-domain is already an internal domain.*

PROOF. Indeed let $\mathbf{A} = (A, \underline{A})$ be any local internal pre-domain and let $\mathbf{f} = (f, \underline{f}) : \mathbf{A} \to \mathbf{A}' = (A', \underline{A}')$ be any topologically conservative morphism into a second internal pre-domain. Then since $A \in \underline{A}_1$ ($\mathbf{A}$ is local) and $f : A \to A'$ is a bijection (being a homeomorphism), we get

$$A' = f[A] \subseteq \underline{f}_1(A) \subseteq A',$$

so that $A' = \underline{f}_1(A) \in \underline{A}_1'$. Thus, $\mathbf{A}'$ is also local and $\mathbf{f}$ dominant. Using the transfer principle, we easily check for any $R \in \underline{A}_n$ that $\underline{f}_n(R) = f_n[R]$. This means that $\mathbf{A}$ may be identified as a point set substructure of $\mathbf{A}'$ and $\mathbf{f}$ with the inclusion morphism. Thus $A = A'$ and for all n, $\underline{A}_n \subseteq \underline{A}_n'$. Our aim is to show the latter inclusions are equalities. Since $\mathbf{A}$ and $\mathbf{A}'$ are local internal pre-domains, there exist dominant morphisms

$$* : {}^{\circ}\mathbf{A} = ({}^{\circ}A, {}^{\circ}\underline{A}) \to \mathbf{A}$$
$$*' : {}^{\circ}\mathbf{A}' = ({}^{\circ}A', {}^{\circ}\underline{A}') \to \mathbf{A}'$$

where ${}^{\circ}\mathbf{A}$ and ${}^{\circ}\mathbf{A}'$ are local full point set structures and the associated maps

$$*_1 : {}^{\circ}\underline{A}_1 \to \underline{A}_1 \text{ and } *_1' : {}^{\circ}\underline{A}_1' \to \underline{A}_1'$$

are bijections. As ${}^{\circ}\mathbf{A}$ and ${}^{\circ}\mathbf{A}'$ are both local and full, it follows easily that for all n,

$$ {}^{\circ}\underline{A}_n = P({}^{\circ}A^n) \text{ and } {}^{\circ}\underline{A}_n' = P({}^{\circ}A'^n).$$

We can identify both ${}^{\circ}A$ and ${}^{\circ}A'$ as subsets of A. But then (routine argument)

$$ {}^{\circ}A = \{p \in A : \{p\} \text{ is } \mathbf{A}\text{-clopen}\} \text{ and }$$
$$ {}^{\circ}A' = \{p \in A : \{p\} \text{ is } \mathbf{A}'\text{-clopen}\},$$

and since the $\mathbf{A}$-topologies *are* the $\mathbf{A}'$-topologies, we see that ${}^{\circ}A = {}^{\circ}A'$ and indeed, that ${}^{\circ}\mathbf{A} = {}^{\circ}\mathbf{A}'$. Now let $R' \in \underline{A}_n'$ be arbitrary. Then an $R \in \underline{A}_n \subseteq \underline{A}_n'$ will exist for which $R' \cap {}^{\circ}A^n = R \cap {}^{\circ}A^n$, and by the transfer principle we get that $R' = R$, so that $R' \in \underline{A}_n$. Thus, for all n, $\underline{A}_n = \underline{A}_n'$ holds, so that indeed $\mathbf{A} = \mathbf{A}'$. Therefore, $\mathbf{f}$ was an isomorphism and $\mathbf{A}$ is an actual (local) internal domain. $\square$

---

[5]The last of these will have immediate application in the Robinson version of "non"standard mathematics.

PROPOSITION 2.7. *Let* $\mathbf{A} = (A, \underline{A})$ *be a local domain with countable cover of* A *by compact clopens, one of which is infinite. Then each* A^n *is compact.*

PROOF. Let $* : {}^\circ\mathbf{A} \to \mathbf{A}$ be the presumed morphism from a full structure (which is clearly local) and let $A = \bigcup_{k=0}^{\infty} A_k$ be the cover by compact clopens. Clearly, compactness impies $A_k \in \underline{A}_1$ for each k. We can assume A_0 is infinite and for every k, that $A_k \subseteq A_{k+1}$. We can write each A_k as $*_1({}^\circ A_k)$ for some ${}^\circ A_k \subseteq {}^\circ A$. Transfer now inplies that ${}^\circ A_0$ is infinite and for every k, that ${}^\circ A_k \subseteq {}^\circ A_{k+1}$.

Suppose now that A is not compact. Then a mapping ${}^\circ f : {}^\circ A \to {}^\circ A$ can be chosen so that for all k, ${}^\circ f[{}^\circ A_0] \not\subseteq {}^\circ A_k$. By transfer, and ${}^\circ\mathbf{A}$'s being full and local, we get $f \in F_{1,1}(\mathbf{A})$ such that $f[A_0] \not\subseteq A_k$ for every k. Since, by Proposition 2.4, f is continuous, we have an open cover of A_0 by the $f^{-1}[A_k]$'s and this cover has no finite subcover. This is impossible, since A_0 is compact.

Thus A is compact. Since ${}^\circ\mathbf{A}$ is full and local with ${}^\circ A$ infinite, there exists surjection ${}^\circ g : {}^\circ A \to {}^\circ A^n$ and hence by transfer a surjection $g : A \to A^n$. By Proposition 2.4, A^n is the continuous image of a compact space, and hence, itself is compact. □

PROPOSITION 2.8. *An internal pre-domain* $\mathbf{A}$ *with a countable cover by compact clopens, one of which is infinite, is already an internal domain.*

PROOF. Assume hypotheses and let $\mathbf{A} \subseteq \mathbf{A}'$ be a topologically conservative inclusion of pre-domains. We need to show that the presumed inclusions $\underline{A}_n \subseteq \underline{A}'_n$ are equalities. Let $\widehat{R}' \in \underline{A}'_n$ be arbitrary and choose local point set structure $\widehat{\mathbf{A}}' = (\widehat{A}', \underline{\widehat{A}}') \subseteq \mathbf{A}'$ which is full in $\mathbf{A}'$ such that also $R' \subseteq \widehat{A}'^n$ and $\widehat{A}'$ contains an infinite compact $\mathbf{A}'$-clopen. By Proposition 2.5, the $\widehat{\mathbf{A}}'$-topologies are the $\mathbf{A}'$-subspace topologies. Clearly $\widehat{\mathbf{A}}'$ is a local internal pre-domain. Since the hypotheses of Proposition 2.7 hold, we have that $\widehat{A}'^n$, and hence, R' is compact. It follows that R' is the union of finitely many elements of $\underline{A}_n$, and hence, itself belongs to $\underline{A}_n$. □

We shall see subsequently that the internal domains, as point set structures, are exactly determined by their topologies and that proper point set morphisms between them are similarly determined.

CHAPTER 3

"Non"standard Analysis

Virtually the single handed invention of Abraham Robinson in the early 1960's (see [**20**]), "non"standard analysis was (among many other things) the three centuries awaited answer as to how to make rigorous sense of Leibniz's original intuitions concerning the new calculus of the 17th century which he had helped to discover. Unlike Newton, Leibniz freely accepted the introduction of infinitely small and infinitely large quantities as ideal elements which would behave (obey the same laws) exactly as ordinary numbers. Leibniz viewed the "existence" of these ideal elements as facilitating mathematical discovery.

Robinson saw that Leibniz's extended mathematical universe, which preserved the same qualities as the "standard" one, corresponded to the model theorist's concept of *elementary extension,* and that the existence of ideal elements in the enlarged domain of discourse also corresponded to the model theorist's notion of *saturation* . Since the twentieth century mathematician's universe has usually been set theoretic, Robinson blended set theory, elementary extensions and saturation to create his "non"standard analysis.

A prevailing prejudice among mathematicians (or logicians speaking for them) has been that one's set theoretic universe should obey the "axiom of regularity" which rules out the existence of infinite $\in$-chains of sets $x_1 \ni x_2 \ni \ldots \ni x_n \ni \ldots$[1], although arbitrarily long chains of this sort, if finite, are acceptable and indeed neccessary in any reasonable set theory. But in the presence of even the mildest saturation, the existence of the latter forces the existence of the former.

For this reason, Robinson chose a compromise which was to blend elementary extensions and saturation with only that fragment of set theory from which an extension of mathematical discourse would leave the axiom of regularity intact. This fragment is modeled by what Robinson termed a *superstructure* . This can be described with the concepts we have previously developed by saying that a superstructure is any full point set structure $\mathbf{X} = (X, \underline{X})$ such that $\underline{X}_1$ consists

---

[1]Part of the attraction of this axiom has been its conformity with the philosophical notion that sets are collections, and, although collections may be of collections, they must ultimately be of things (urlments) which are not collections. It is possible (and perhaps useful) to adopt a more agnostic idea of sets.

of all possible sets X' for which there exists a single n so that every possible $\in$-chain $x_1 \ni x_2 \ni \ldots \ni x_k$ with first entry $x_1 \in X'$ will have, if continued sufficiently long, an entry $x_i \in X$ for some $i \leq n$ for which $x_i \cap X = \emptyset$. Letting X_0 consist of all $x \in X$ for which $x \cap X = \emptyset \neq x$, the superstructure $\mathbf{X}$ can be alternatively identified as

$$X = \bigcup_{k \geq 0} X_k \text{ with } X_{k+1} = \mathcal{P}(\bigcup_{k \geq e} X_e)$$
$$\underline{X}_n = \bigcup_{k \geq 0} \mathcal{P}(X_k^n).$$

Notice that if one encodes ordered n-tuples as sets in the usual manner, namely

$$(x, y) = \{\{x\}, \{x, y\}\} \text{ and } (x_1, x_2, \ldots, x_{k+1}) = ((x_1, x_2, \ldots, x_k), x_{k+1}),$$

then each $\mathbf{X}$-basis set or relation is actually an element of X. Also clear is that for each $\mathbf{X}$-basis set X' that $\in |_{X'} = \{(x, y) \in X'^2 : x \in y\}$ is an $\mathbf{X}$-basis binary relation. Elements $x \in X_0$ are called *individuals* of the superstructure. General elements $x \in X$ are called *entities*.

Robinson's central concept in his "non"standard analysis is of the "enlargement" of a superstructure. We might call a *morphism* $\mathbf{X} \to \mathbf{X}'$ of superstructures any morphism $\mathbf{f} = (f, \underline{f})$ of the underlying point set structures $\mathbf{X} = (X, \underline{X})$ and $\mathbf{X}' = (X', \underline{X}')$ for which

$$f[X_0] \subseteq X_0' \text{ (preservation of individuals)},$$
$$\text{for } R \in \underline{X}_n, \underline{f}_n(R) = f(R) \text{ (recall } R \in X),$$
$$f(X_0) = X_0', \text{ and for } X'' \in \underline{X}_1, f(\in |_{X''}) = \in |_{f(X'')}.$$

Clearly, the entire morphism $\mathbf{f}$ can be identified with f, which Robinson writes as $*$ and also writes $*x$ for $*(x)$. We shall ourselves write $*\mathbf{X}$ for the image point set substructure $*[\mathbf{X}] \subseteq \mathbf{X}'$ corresponding to $* : \mathbf{X} \to \mathbf{X}'$. For Robinson the morphism $* : \mathbf{X} \to \mathbf{X}'$ is *enlarging* (and $\mathbf{X}'$ is an *enlargement* of $\mathbf{X}$) if $*\mathbf{X}$ is 1-saturated or, in other words, if all $*\mathbf{X}$-basis sets and relations are compact in the $*\mathbf{X}$-topologies. The enlargement $\mathbf{X}'$ is κ-*saturated* if the point set structure $*\mathbf{X}$ is. We shall routinely assume (it is the only case of interest) that the set of individuals X_0 involved in a Robinson enlargement $\mathbf{X} \to \mathbf{X}'$ is always infinite.

Since $\mathbf{X}$ is full and the resulting point set morphism $* : \mathbf{X} \to *\mathbf{X}$ is dominant, it is clear that $*\mathbf{X}$ is an internal pre-domain in the sense defined earlier. Since $*X$ has a countable cover by the increasing clopen compact $*\mathbf{X}$-basis sets $*X_k$ ($k \geq 0$), it follows from Proposition 2.8 that $*\mathbf{X}$ is an internal domain. Robinson calls the entities $y \in *X$ *internal* and if a particular example is of the form $y = *x$ for $x \in X$, then y is called the *standard copy* of the *standard* element x. Entities $y \in X'$ which are not internal are called *external*. It is routine to identify X_0 as a subset of X_0' via the map $*$ so that each individual x in $\mathbf{X}$ is its own standard copy in $\mathbf{X}'$.

Of course the *transfer principle* (Theorem 2.1) holds in this context: Each local sentence φ about standard entities in $\mathbf{X}$ (Robinson calls such sentences "bounded") has an interpretaion $*\varphi$ as a local sentence about their standard copies in $\mathbf{X}'$, and φ is true exactly when $*\varphi$ is true.

Let us see how Leibniz gets his infinitesimal numbers in a Robinson enlargement $* : \mathbf{X} \to \mathbf{X}'$. We can assume the original superstructure $\mathbf{X}$ contains amongst its individuals the full set R of real numbers. With our identifications, we have $R \subseteq {}^*R$ and, by the transfer principle, all operations on R canonically extend to similar operations on ${}^*R$ having the same properties, so that ${}^*R$ is naturally an ordered field extending R. Subsequent authors have called this the field of *hyperreal* numbers. For positive natural numbers n, the open intervals $(0, 1/n)$ satisfy the finite intersection property and thus, so do their standard copies ${}^*(0, 1/n) \subseteq {}^*R$. But these sets are clopens in the ${}^*\mathbf{X}$-topology on ${}^*R$ which is compact. Thus there exists a hyperreal $dx \in {}^*R$ satisfying

$$0 < dx < 1/n \text{ for all positive natural n.}$$

Similar techniques construct infinite positive and negative hyperreal numbers. Infinite hypernatural numbers exist and, by the transfer principle, each has a unique prime factorization (by possibly infinite primes and infinitely many of them if necessary). In such manner is mathematical mythology rendered respectable.

A subtle point concerning the transfer principle can be mentioned. The original field of reals R is "archimedean""in the sense that

> for every real $x \in R$ there exists a natural
> number n for which $\mid x \mid < n$.

Clearly, the hyperreals (having infinite numbers) fail this principle, but does this not violate the transfer principle? The reason it doesn't is that in the $*$-version of the archimedean property the quantifier "there exists a natural number n..." becomes "there exists a *hyper*natural number n...". In this sense, the hyperreals are not archimedean, but are $*$-archimedean. The *set* of standard natural numbers N, considered as a subset of the set ${}^*N$ of hypernatural numbers, turns out to be a *non*internal set. It is an early example of a truly external set.

As an illustration of "non"standard mathematics in action, I offer the following "non"standard proof of a well known result in model theory, called the "compactness theorem". This states that if you have a theory T and if every finite set $T_0 \subseteq T$ of its theorems has a model, then T itself has a model. The standard proofs are quite straightforward as such, but they all fail to link actual topology to that suggested in the theorem's name.

A "non"standard proof: Let $\mathbf{L}$ be the language of T. Since any $\mathbf{L}$-structure $\mathbf{A} = (A, \dots)$ has an elementary substructure $\mathbf{A}' = (A', \dots) \subseteq \mathbf{A}$ whose universe A' is "small", we can choose a set *Struct* of $\mathbf{L}$-structures which contains all possible elementary equivalence types. For each sentence φ of $\mathbf{L}$, let

$$MOD_\varphi = \{\mathbf{A} \in Struct : \mathbf{A} \models \varphi\}.$$

Our assumptions concerning T amount to saying that the family

$$\{MOD\varphi\}_{\varphi \in T}$$

of subsets in $Struct$ satisfies the finite intersection property.

Assume now that all the foregoing sets and their related collections have been encoded as standard sets in some superstructure $\mathbf{X}$ and that we have chosen an enlargement $* : \mathbf{X} \rightarrow \mathbf{X}'$. By the transfer principle, the standard copy $*\mathbf{L}$ of $\mathbf{L}$ acts like a language extending $\mathbf{L}$. However, included in $*\mathbf{L}$ will possibly be relation or operator symbols whose "arities" are infinite *hyper*natural numbers. Formulas in $*\mathbf{L}$ may also have infinite (although hyperfinite) length, since hyperfinite conjunctions, disjunctions, etc. are permitted. The transfer principle further shows that the elements $\mathbf{A}' \in Struct$ act as $*\mathbf{L}$-structures and if we let

$$\models = \{(\mathbf{A}, \varphi) : \mathbf{A} \in Struct, \ \varphi \text{ an } \mathbf{L}\text{-sentence}, \ \mathbf{A} \models \varphi\},$$

then $*\models$ acts as a satisfaction relation between $*\mathbf{L}$-structures $\mathbf{A}' \in *Struct$ and $*\mathbf{L}$-sentences φ'.

So far we have only been moving around in the internal part $*\mathbf{X} \subseteq \mathbf{X}'$ of the enlargement $* : \mathbf{X} \rightarrow \mathbf{X}'$. We now do an external construction which is typical in "non"standard mathematics. Given any $*\mathbf{L}$-structure $\mathbf{A}' \in *Struct$ we may remove all of its items (constants, functions, relations, etc.) which don't pertain to the language $\mathbf{L}$ and derive a *new* $\mathbf{L}$-structure. Let us call this the *standard part* of $\mathbf{A}'$, and write it as $^{st}\mathbf{A}'$. It is a strictly external entity in the enlargement $\mathbf{X}'$. An inductive argument shows for any sentence φ of $\mathbf{L}$ that

$$\mathbf{A}' \ ^* \models \ ^*\varphi \Leftrightarrow \ ^{st}\mathbf{A}' \models \varphi.$$

We are done. The collection $\{^*MOD\varphi\}_{\varphi \in T}$ is a family of closed subsets in a compact space $*Struct$ which satisfies the finite intersection property. Therefore, the family has a nonempty intersection and, if we pick

$$\mathbf{A}' \in \bigcap_{\varphi \in T} \ ^*MOD\varphi,$$

then $^{st}\mathbf{A}'$ is a model of T.

Although the specialist will recognize the foregoing proof as valid, I have intentionally omitted many details. As a test of understanding, both of model theory and of "non"standard analysis, I strongly urge the more general reader to work these details out. A good general reference for Robinson-style "non"standard mathematics is [12]. Other texts of related interest are [7], [9], [16], [17], [18], [19], [20], [21], [22] and [23].

Part 2

Topological Aspects

CHAPTER 4

Introduction

If $\mathbf{X}$ is a superstructure with enlargement $* : \mathbf{X} \to \mathbf{X}'$ there is, as discussed earlier, an image point set substructure $^*\mathbf{X} \subseteq \mathbf{X}'$ made up of all the internal entities of the enlargement. We call $^*\mathbf{X}$ the *internal part* of the enlargement. The enlarging morphism $* : \mathbf{X} \to \mathbf{X}'$ factors as a morphism of point set structures $* : \mathbf{X} \to {}^*\mathbf{X} \subseteq \mathbf{X}'$. Assuming (as always) that the set of individuals X_0 is infinite, we have seen that $^*\mathbf{X}$ is an internal domain. Indeed, it is a *non*local internal domain.

With any point set structure one can study the extent to which topology determines its structure. It turns out that every internal domain is exactly determined by its topology. I shall present the details of this story in Part 2.

We begin with a basic initial observation.

PROPOSITION 4.1. *Let* $\mathbf{A} = (A, \underline{A})$ *be an arbitrary internal domain with morphism* $* = (*, \underline{*}) : {}^\circ\mathbf{A} \to \mathbf{A}$ *from a full structure* ${}^\circ\mathbf{A} = ({}^\circ A, {}^\circ\underline{A})$ *for which the associated map* $\underline{*}_1 : {}^\circ\underline{A}_1 \to \underline{A}_1$ *is a bijection. Then the following are true:*

    a) $*[{}^\circ A] = \{p \in A : \{p\} \subseteq A$ *is* $\mathbf{A}$-*clopen*$\} = \{p \in A : p$ *is* $\mathbf{A}$-*isolated*$\}$

    b) *the morphism* $* : {}^\circ\mathbf{A} \to \mathbf{A}$ *is proper and dominant.*

PROOF. Clearly for $p \in {}^\circ A$, p is an individual in ${}^\circ\mathbf{A}$, hence $*(p)$ is also in $\mathbf{A}$, so that $\{*(p)\} \subseteq A$ is $\mathbf{A}$-clopen. Conversely, if $\{p'\} \subseteq A$ is $\mathbf{A}$-clopen then $\{p'\} = \underline{*}_1(S)$ for some $S \in {}^\circ\underline{A}_1$. By transfer, $\emptyset \neq S$ and if $p \in S$, then $*(p) \in \{p'\}$, so $*(p) = p'$. This forces $S = \{p\}$ and thus that $p' \in *[{}^\circ A]$. This shows (a).

To argue (b), we need to show that each associated map $\underline{*}_n : {}^\circ\underline{A}_n \to \underline{A}_n$ is a bijection. By the transfer principle we know it is one to one. To each relation $R \in \underline{A}_n$, assign the relation ${}^\circ R = \underline{*}_n^{-1}[R]$. Since $R \subseteq A'^n$ for some $A' \in \underline{A}_1$, it is clear that ${}^\circ R \in {}^\circ\underline{A}_n$. To show (b), we need only argue that in every case $\underline{*}_n({}^\circ R) = R$.

If $\underline{*}_n({}^\circ R) = R$, fails then ${}^\circ A'$ is forced to be infinite (since otherwise, by transfer, $*[{}^\circ A'] = A'$). Thus, there is a bijection ${}^\circ h : {}^\circ A'^n \to {}^\circ A'$ whose graph ${}^\circ G$ is an element of ${}^\circ\underline{A}_{n+1}$. Transfer implies that $G = \underline{*}_{n+1}({}^\circ G)$ is the graph of a bijection $h : A'^n \to A'$ for which $h \circ *_n|_{{}^\circ A'^n} = *_1 \circ {}^\circ h$. By definition, for any relation $R' \subseteq A'^n$ we have

$$R' \in \underline{A}_n \Leftrightarrow h[R'] \in \underline{A}_1.$$

By injections and bijections, we have

$$^{\circ}(h[R']) = {}^{\circ}h[{}^{\circ}R']$$

and, by transfer, also that

$$h[\underline{*}_n({}^{\circ}R')] = \underline{*}_1({}^{\circ}h[{}^{\circ}R']).$$

From $\underline{*}_n({}^{\circ}R) \neq R$, we get

$$\underline{*}_1({}^{\circ}(h[R])) = \underline{*}_1({}^{\circ}h[{}^{\circ}R]) = h[\underline{*}_n({}^{\circ}R)] \neq h[R]$$

and thus, for the counterexample $\underline{*}_n({}^{\circ}R) \neq R$, we can assume that $n = 1$. But $\underline{*}_1 : {}^{\circ}\underline{A}_1 \to \underline{A}_1$ *is* a bijection, so $R = \underline{*}_1(S)$ for some $S \subseteq {}^{\circ}A'$. Thus,

$$\underline{*}_1({}^{\circ}(\underline{*}_1(S))) = \underline{*}_1({}^{\circ}R) \neq R = \underline{*}_1(S) \,,$$

which implies that $^{\circ}(\underline{*}_1(S)) \neq S$. But by transfer, $^{*}({}^{\circ}A' - S) = A' - {}^{*}(S)$ so that

$$S \subseteq {}^{\circ}(\underline{*}_1(S)) \text{ and } {}^{\circ}A' - S \subseteq {}^{\circ}(\underline{*}_1({}^{\circ}A' - S)) = {}^{\circ}A' - {}^{\circ}(\underline{*}_1(S)) \,,$$

which contradicts $^{\circ}(\underline{*}_1(S)) \neq S$. This proves (b). $\square$

The message here is that in effect the **A**-topologies exactly reproduce $^{\circ}\mathbf{A}$ from **A** and also determine the morphism $^{\circ}\mathbf{A} \to \mathbf{A}$ completely. Indeed, if we use $*$ to identify $^{\circ}A$ as a subset of A, then

$$^{\circ}A = \{p \in A : \{p\} \subseteq A \text{ is } \mathbf{A}\text{-clopen}\}$$
$$^{\circ}\underline{A}_n = \{R \cap {}^{\circ}A^n : R \in \underline{A}_n\}$$
$$\text{for } {}^{\circ}R \in {}^{\circ}\underline{A}_n, \ \underline{*}_n({}^{\circ}R) = \overline{{}^{\circ}R} \quad (\mathbf{A}\text{-topologies}).$$

Clearly, topological determinacy is afoot. We shall finish the story of this topological determinacy in three stages. In Chapter 5, we study the characteristic topologies which **A** imposes on each of its A^n's. I term these *CL spaces* . In Chapter 6, we describe the topological determinacy when **A** is a *local* internal domain. In Chapter 7, we complete the description of topological determinacy when **A** is an arbitrary internal domain.

Theory of CL Spaces

The **A**-topologies which an internal domain **A** imposes on its finitary powers A^n all share certain fundamental characteristics. Each is zero dimensional (clopen sets form a basis), for each its subset $^\circ A^n$ of isolated points is dense, and for each the closure of any subset in this $^\circ A^n$ is an **A**-clopen. This motivates our opening definition:

DEFINITION 5.1. *A topological space X is a CL space if it is zero dimensional, its subset $^\circ X$ of isolated points is dense, and the closure of each subset in $^\circ X$ is clopen. The* core *of a CL space is the set $^\circ X$ of its isolated points. When X is a CL space and $W \subseteq X$ is a subset, we will write $^\circ W$ for $W \cap {}^\circ X$.*

An easy example of a CL space is the Stone-Čech compactification $X = \beta(^\circ X)$ of a set $^\circ X$ with its discrete topology (all subsets open), but this has the uncharacteristic feature of being Hausdorff. In fact, up to homeomorphism, any Hausdorff CL space X is just a dense subspace of $\beta(^\circ X)$. The problem with Stone-Čech compactification is that as an operation it fails to commute with cartesian products. There is no natural way — not even set theoretically — to identify $\beta(^\circ X \times {}^\circ X)$ as $\beta(^\circ X) \times \beta(^\circ X)$. For this reason (as warned previously), our investigations will require *non*Hausdorff CL spaces. Typically, no point in $X - {}^\circ X$ will be closed.

Our opening proposition shows that being CL is a local property.

PROPOSITION 5.1. *Assume X is a topological space with a clopen cover by CL subspaces. Then X itself is a CL space.*

PROOF. Of the three conditions required in the definition of a CL space, only the third is not immediate. Let $Z \subseteq {}^\circ X$ be arbitrary. We need only show $\overline{Z} \subseteq X$ is clopen. Let $Z' = {}^\circ X - Z$ and let $X = \bigcup X_i$ be the clopen cover of X by CL subspaces. Clearly, $\overline{Z}$ and $\overline{Z}'$ cover X since for each i

$$X_i = \overline{Z \cap X_i} \cup \overline{Z' \cap X_i} \subseteq \overline{Z} \cup \overline{Z}'.$$

Also, if $\overline{Z} \cap \overline{Z}' \neq \emptyset$ then for some i

$$\overline{Z \cap X_i} \cap \overline{Z' \cap X_i} = \quad (X_i \subseteq X \text{ clopen})$$
$$(\overline{Z} \cap X_i) \cap (\overline{Z}' \cap X_i) = \overline{Z} \cap \overline{Z}' \cap X_i \neq \emptyset.$$

However, $Z' \cap X_i$ is an open set disjoint from $Z \cap X_i$, and so is disjoint from $\overline{Z \cap X_i}$. But $\overline{Z \cap X_i}$ is clopen and, being disjoint from $Z' \cap X_i$, would have to be disjoint from $\overline{Z' \cap X_i}$. Thus $\overline{Z} \cap \overline{Z}' \neq \emptyset$ leads to a contradiction. Therefore $\overline{Z}$ and $\overline{Z}'$ partition X, and $\overline{Z}$ is clopen. $\square$

The next proposition shows the CL spaces carry a sort of rudimentary "transfer principle".

PROPOSITION 5.2. *Assume X is a CL space. Then*
   a) *$W \subseteq X$ is clopen $\Leftrightarrow W = {}^{\circ}\overline{W}$*
   b) *For any $Z, Z' \subseteq {}^{\circ}X$ one has*
$${}^{\circ}(\overline{Z}) = Z, \ \overline{Z} \cap \overline{Z}' = \overline{Z \cap Z'}, \ \overline{Z} \cup \overline{Z}' = \overline{Z \cup Z'},$$
$$\overline{Z} - \overline{Z}' = \overline{Z - Z'} \ \text{and} \ \overline{Z} = \overline{Z}' \Leftrightarrow Z = Z'.$$

PROOF. For (a), the right to left implication follows from definitions. Conversely, assume $W \subseteq X$ is clopen. Then trivially ${}^{\circ}\overline{W} \subseteq W$. If $p \in W - {}^{\circ}\overline{W}$ existed, then $p \in {}^{\circ}X$ can be assumed ($W - {}^{\circ}\overline{W}$ is open, ${}^{\circ}X$ is dense) which forces $p \in W \cap {}^{\circ}X = {}^{\circ}W$ — a contradiction. For (b), everything follows easily from (a) and (b)'s first item ${}^{\circ}(\overline{Z}) = Z$. Clearly, $Z \subseteq {}^{\circ}(\overline{Z})$ since $Z \subseteq {}^{\circ}X$ is assumed. If $p \in {}^{\circ}(\overline{Z}) - Z$ were to exist, then $p \in \{p\}$ (an open set) and $\{p\} \cap Z = \emptyset$ and hence, $p \notin \overline{Z}$ — a contradiction. $\square$

It is natural to distinguish two types of maps between CL spaces.

DEFINITION 5.2. *A map $f : X \to Y$ between CL spaces is a CL_0 map if it is continuous and preserves clopen points; i.e., if $f[{}^{\circ}X] \subseteq {}^{\circ}Y$. It is a CL map if it is continuous and preserves arbitrary clopen sets; i.e., $f[Z] \subseteq Y$ is clopen for each clopen $Z \subseteq X$. When $f : X \to Y$ is a CL_0 map between CL spaces, we write ${}^{\circ}f$ for the induced map ${}^{\circ}X \to {}^{\circ}Y$ between the cores.*

We complete this section with a brief exploration of the properties of CL_0 and CL maps.

PROPOSITION 5.3. *Assume $f : X \to Y$ is a CL_0 mapping between CL spaces. Then*
   a) *for any $W \subseteq {}^{\circ}Y$, $f^{-1}[\overline{W}] = \overline{{}^{\circ}f^{-1}[W]}$*
   b) *f is a CL mapping $\Leftrightarrow$ for all $Z \subseteq {}^{\circ}X$, $\overline{f[Z]} \subseteq f[\overline{Z}]$*
   c) *if f is a bijection, then both f and f^{-1} are CL maps*
   d) *if f is $(1-1)$, then it is an immersion.*

PROOF. (a) follows from Proposition 5.2a, the fact that $f^{-1}[\overline{W}]$ must be clopen, and the assumption that f is a CL_0 map. For (b), the right to left implication follows from definitions. Suppose, conversely, that f is a CL map and that $Z \subseteq {}^\circ X$. Then $\overline{f[Z]} \subseteq f[\overline{Z}]$ since $f[Z] \subseteq f[\overline{Z}]$ and $f[\overline{Z}]$ is clopen. For (c), assume f is $(1-1)$ and onto. By (a), f^{-1} is a $(1-1)$ onto map $X \leftarrow Y$ preserving clopen sets and thus, by (b), is CL if continuous. For $Z \subseteq {}^\circ X$,

$$(f^{-1})^{-1}[\overline{Z}] = f[\overline{Z}] =$$
$$f[{}^\circ f^{-1}[{}^\circ f[Z]] = \underline{\text{(use (a))}}$$
$$f[f^{-1}[{}^\circ f[Z]]] = {}^\circ f[Z]$$

which is clopen in Y. Thus f^{-1} *is* continuous and hence CL and, by duality, $f = (f^{-1})^{-1}$ is also CL. Finally, (d) follows from (c) and the observation that $f[X] \subseteq Y$, with its subspace topology, is also a CL space with core $f[{}^\circ X]$. □

PROPOSITION 5.4. *A map* $f : X \to Y$ *between* CL *spaces is* CL *if and only if* $f[{}^\circ X] \subseteq {}^\circ Y$ *and* $f[\overline{Z}] = \overline{f[Z]}$ *for all* $Z \subseteq {}^\circ X$.

PROOF. The left to right implication is clear from Proposition 5.3b. Suppose, conversely, that $f : X \to Y$ is an arbitrary mapping between CL spaces such that $f[{}^\circ X] \subseteq {}^\circ Y$ and for any $Z \subseteq {}^\circ X$, $f[\overline{Z}] = \overline{f[Z]}$ holds. Since $f[{}^\circ X] \subseteq {}^\circ Y$, to show f is CL we need only verify that it is continuous. Let $W \subseteq {}^\circ Y$ be arbitrary. Then $\overline{W} \subseteq Y$ is a typical basis open set and, if $W' = {}^\circ Y - W$, then $\overline{W}$, $\overline{W}'$ form (Proposition 5.2b) a clopen partition of Y. Let $Z = {}^\circ f^{-1}[W]$ and $Z' = {}^\circ f^{-1}[W']$. Then Z and Z' form a partition of ${}^\circ X$, hence $\overline{Z}$ and $\overline{Z}'$ form a clopen partition of X. Since

$$f[\overline{Z}] = \overline{f[Z]} = \overline{{}^\circ f[Z]} \subseteq \overline{W}$$

and similarly $f[\overline{Z}'] \subseteq \overline{W}'$, one has that $\overline{Z} = f^{-1}[\overline{W}]$ and f is seen to be continuous. □

CHAPTER 6

Topological Determinacy of Local Internal Domains

If $\mathbf{A} = (A, \underline{A})$ is a local internal domain, then all the finitary $\mathbf{A}$-topologies are CL spaces. Additionally, it is clear that each of the projection maps $A^n \to A^m$ is CL. In this section we shall describe and demonstrate how all local internal domains and proper point set morphisms between them are exactly determined by their topologies.

We recall the following notation: When $f, g : X \to X$ are maps, we write $f \times g : X \times X \to X \times X$ for the map sending $< p, q >$ to $< f(p), g(q) >$. We also write $< f, g >: X \to X \times X$ for the map sending p to $< f(p), g(p) >$. We use $G(f) \subseteq X \times X$ to denote the graph of f. For any set W, we let $\Delta_W = \{< p, p >: p \in W\} \subseteq W \times W$ be the *diagonal set* of W.

Our first goal is the proof of the following theorem.

THEOREM 6.1. *Let* $\mathbf{A} = (A, \underline{A})$ *be a local internal domain, and let* $F(A)$ *denote the collection of all maps* $h : A \to A$ *for which the graph* $G(h) \subseteq A^2$ *is clopen. Then the* $\mathbf{A}$-*topologies on* A *and* A^2 *and the function family* $F(A)$ *satisfy the following properties:*

1) *A and A^2 are both CL spaces*
2) *both projection maps $\pi_i : A^2 \to A$ (i=1,2) are CL maps*
3) *$F(A)$ contains the identity map $id : A \to A$*
4) *for any $h, k \in F(A)$, $h \times k : A^2 \to A^2$ is a CL map*
5) *A^2 is covered by sets of the form $< h, k > [A]$ for $h, k \in F(A)$.*

Conversely, if A is any nonempty set and topologies on A and A^2 are given which satisfy the foregoing properties, then there exists a unique $\underline{A}$ such that $\mathbf{A} = (A, \underline{A})$ is a local internal domain whose $\mathbf{A}$-topologies on A and A^2 coincide with those given.

I shall proceed the proof of Theorem 6.1 with a series of lemmas. For the sake of discussion, we shall informally call any set A with topologies on A and A^2 satisfying properties (1) through (5) listed above a *topological* local internal domain.

LEMMA 6.1. *Let A be a topological local internal domain. Then*
  a) $°(A \times A) = °A \times °A$
  b) *for $Z, W \subseteq °A$, $\overline{Z \times W} = \overline{Z} \times \overline{W}$*
  c) *for any $f, g \in F(A)$, $<f,g>: A \to A \times A$ is CL*
  d) *for any $f \in F(A)$, $f : A \to A$ is CL*
  e) *for any $f, g \in F(A)$, $f \circ g \in F(A)$.*

PROOF. For (a): since $\pi_i : A \times A \to A$ $i = 1, 2$ are CL, $\pi_i[°(A \times A)] \subseteq °A$ for $i = 1, 2$ and hence, $°(A \times A) \subseteq °A \times °A$. Also for $p, q \in °A$,

$$\{<p,q>\} = \pi_1^{-1}[\{p\}] \cap \pi_2^{-1}[\{q\}],$$

which is an intersection of clopen sets; thus, $<p,q> \in °(A \times A)$, i.e., $°A \times °A \subseteq °(A \times A)$. Part (b) follows from observing that

$$\overline{Z} \times \overline{W} = \pi_1^{-1}[\overline{Z}] \cap \pi_2^{-1}[\overline{W}],$$

which is an intersection of clopen sets, hence itself is clopen, and observing that $(\overline{Z} \times \overline{W}) \cap (°A \times °A) = Z \times W$. To prove (c), note for $f, g \in F(A)$ that

$$<f,g> = (f \times g) \circ <id, id>,$$

so that it suffices to show that $<id, id>: A \to A \times A$ is CL. Clearly,

$$<id, id> [°A] \subseteq °A \times °A.$$

Also, for $Z \subseteq °A$, remembering that $\Delta_A = G(id)$ is clopen, we have

$$<id, id> [\overline{Z}] = \Delta_{\overline{Z}} = \Delta_A \cap (\overline{Z} \times A) = \overline{\Delta_{°A}} \cap \overline{(Z \times °A)}$$
$$= \overline{\Delta_{°A} \cap (Z \times °A)} = \overline{\Delta_Z} = \overline{<id, id> [Z]}.$$

So by Proposition 5.4, $<id, id>$ is a CL map. Part (d) follows from (c) and that fact that $f = \pi_2 \circ <id, f>$. To prove (e), it suffices to show that $<id, f \circ g>: A \to A \times A$ is a CL map, since $G(f \circ g) = <id, f \circ g> [A]$. But

$$<id, f \circ g> = (id \times f) \circ <id, g>$$

and thus, is a composition of CL maps. $\square$

LEMMA 6.2. *Let A be a topological local internal domain. Then for any map $°h : °A \to °A$ there exists a unique $h \in F(A)$ such that $h|_{°A} = °h$.*

PROOF. Suppose such $h \in F(A)$ existed. Then

$$G(h) = \overline{°G(h)} = \overline{G(°h)},$$

so that h must be unique. For the existence of h, it suffices to show that $G = \overline{G(°h)}$ is the graph of a map $h : X \to X$. Since

$$\pi_1[G] = \pi_1[\overline{G(°h)}] = \overline{\pi_1[G(°h)]} = \overline{°A} = A,$$

we have that for each $p \in A$, there exists $q \in A$ such that $<p, q> \in G$. Suppose next that

$$<p, q>, <p, q'> \in G.$$

Using "axiom" (5) in the definition of topological local internal domains, pick

$$f, f', g, g' \in F(A)$$
$$\text{and } r, r' \in A$$

such that

$$<f, g> (r) = <p, q> \text{ and}$$
$$<f', g'> (r') = <p, q'>.$$

Again, pick $k, k' \in F(A)$ and $t \in A$ such that $<k, k'> (t) = <r, r'>$. Then

$$<fk, gk> (t) = <p, q> \text{ and}$$
$$<f'k', g'k'> (t) = <p, q'>$$

and, by Lemma 6.1e,

$$fk, gk, f'k', g'k' \in F(A),$$

so we may as well use fk in place if $f'k'$ and write f, g, g' in place of $fk, gk, g'k'$, respectively. We have then that

$$<f, g> (t) = <p, q>, \quad <f, g'> (t) = <p, q'>,$$

and hence, that t lies in both of the clopen sets

$$<f, g>^{-1} \overline{[G(\circ h)]}, \quad <f, g'>^{-1} \overline{[G(\circ h)]}.$$

There exists then $Z \subseteq {}^{\circ}A$ such that $t \in \overline{Z}$ and

$$<f, g> [\overline{Z}], \quad <f, g'> [\overline{Z}] \subseteq \overline{G(\circ h)}.$$

Then for all $p' \in Z$, $g(p') = {}^{\circ}h(f(p')) = g'(p')$. Thus $g|_Z = g'|_Z$ and hence,

$$Z \subseteq <g, g'>^{-1} [\Delta_{\circ A}],$$

which implies

$$t \in \overline{Z} \subseteq <g, g'>^{-1} \overline{[\Delta_{\circ A}]} = <g, g'>^{-1} [\Delta_A],$$

which in turn implies

$$q = g(t) = g'(t) = q'.$$

Thus, $G = \overline{G(\circ h)}$ is the graph of a function $h : A \to A$. $\quad\square$

LEMMA 6.3. *Let A be an infinite topological local internal domain. Then there exists $f, g \in F(A)$ such that $<f, g>: A \to A \times A$ is a homeomorphism.*

PROOF. Since A is infinite, $^\circ A$ must be also. Pick $^\circ f, ^\circ g : {}^\circ A \to {}^\circ A$ so that $<{}^\circ f, {}^\circ g>: {}^\circ A \to {}^\circ A \times {}^\circ A$ is a bijection. Choose $f, g \in F(A)$ extending $^\circ f, ^\circ g$. By Proposition 5.3c and Lemma 6.1c, it suffices to show that $<f, g>: A \to A \times A$ is a bijection. It is clearly onto, since

$$<f,g>[A] = <f,g>[^\circ \overline{A}] = \overline{<{}^\circ f, {}^\circ g>[^\circ A]} = \overline{^\circ A \times {}^\circ A} = A \times A.$$

Also

$$\{<p,p'>\in A^2 :<f,g>(p) =<f,g>(p')\}$$
$$= (f \times f)^{-1}[\Delta_A] \cap (g \times g)^{-1}[\Delta_A]$$
$$= \overline{(^\circ f \times {}^\circ f)^{-1}[\Delta_{^\circ A}]} \cap \overline{(^\circ g \times {}^\circ g)^{-1}[\Delta_{^\circ A}]}$$
$$= \overline{(^\circ f \times {}^\circ f)^{-1}[\Delta_{^\circ A}] \cap (^\circ g \times {}^\circ g)^{-1}[\Delta_{^\circ A}]}$$
$$= \overline{\{<p,p'>\in (^\circ A)^2 :<{}^\circ f, {}^\circ g>(p) = <{}^\circ f, {}^\circ g>(p')\}}$$
$$= \overline{\Delta_{^\circ A}} \ \ (<{}^\circ f, {}^\circ g> \text{ is one to one}) \ \ = \Delta_A.$$

Thus, $<f,g>$ is also one to one. $\square$

We can now introduce canonical topologies on the finitary powers A^n of a topological local internal domain A. As before, let $F(A)$ be the collection of maps $f : A \to A$ whose graphs $G(f) \subseteq A \times A$ are clopen. For each $n \geq 1$, give A^n the *canonical topology* using as an open *subbasis* the sets of the form $<f_1, \ldots, f_n>[A]$ where $f_1, \ldots, f_n \in F(A)$. For $n, m \geq 1$, let $F_{n,m}(A)$ be the set of all maps $f : A^n \to A^m$ whose graph $G(f) \subseteq A^{n+m}$ is clopen in the canonical topology.

LEMMA 6.4. *Let A be a topological local internal domain. Then*
   a) *for each $n \geq 1$, the canonical topology turns A^n into a CL space with $^\circ(A^n) = (^\circ A)^n$*
   b) *for $n = 1, 2$, the canonical topologies coincide with the ones already given on A and $A \times A$*
   c) *for any $n, m \geq 1$, the elements of $F_{n,m}(A)$ are CL maps and for $n \geq m$ include all projection and identity maps*
   d) *each map $^\circ f : {}^\circ A^n \to {}^\circ A^m$ extends uniquely to some element $f : A^n \to A^m$ in $F_{n,m}(A)$*
   e) *$g \circ f \in F_{n,k}(A)$ for all $f \in F_{n,m}(A)$ and $g \in F_{m,k}(A)$*
   f) *$<f,g>\in F_{n,m+k}(A)$ for all $f \in F_{n,m}(A)$ and $g \in F_{n,k}(A)$*
   g) *$f \times g \in F_{n+k,m+l}(A)$ for all $f \in F_{n,m}(A)$ and $g \in F_{k,l}(A)$.*

PROOF. The theorem is a triviality when $^\circ A$ is finite, for in this case $^\circ A$ is itself clopen, $A = \overline{^\circ A} = {}^\circ A$, and both A and $A \times A$ have the discrete topologies. We may assume then that $^\circ A$ is infinite. Using Lemma 6.3, pick $f, g \in F(A)$ such that $<f,g>: A \to A \times A$ is a homeomorphism. Define bijections $\tau_n : A \to A^n$ inductively by

$$\tau_1 = id, \ \tau_2 =<f,g>, \text{ and for } n \geq 2,$$
$$\tau_{n+1} = (id_{n-1} \times \tau_2) \circ \tau_n$$

where id_{n-1} is the identity map on A^{n-1}. Topologize each A^n so that τ_n is a homeomorphism from $Dom(\tau_n) = A$ with its original topology. Clearly, in these τ-topologies each A^n is a CL space with core $(^{\circ}A)^n$. Our goal (to prove (a)) is to show these τ-topologies are the canonical ones. An easy induction shows each τ_n can be written in the form $< f_1, \ldots, f_n >$ for certain $f_1, \ldots, f_n \in F(A)$.

Our first task is to show, for any $h_1, \ldots, h_n \in F(A)$, that

$$< h_1, \ldots, h_n >: A \to A^n$$

is a CL map in the τ-topologies. Since $id_{n-1} \times \tau_2 : A^n \to A^{n+1}$ is a τ-homeomorphism, a proof of this by induction on n is possible if we can show, for arbitrary $h, k \in F(A)$, that

$$\tau_2^{-1} \circ < h, k >= < f, g >^{-1} \circ < h, k >: A \to A$$

is an element of $F(A)$. Since $< f, g >$ is bijective we can pick

$$^{\circ}H : {^{\circ}A} \to {^{\circ}A}$$

such that

$$< {^{\circ}f}, {^{\circ}g} > \circ ({^{\circ}H}) = < {^{\circ}h}, {^{\circ}k} >,$$

i.e., $({^{\circ}f}) \circ ({^{\circ}H}) = {^{\circ}h}$ and $({^{\circ}g}) \circ ({^{\circ}H}) = {^{\circ}k}$. Let H be the unique extension in $F(A)$ of $^{\circ}H$. Then in the original topologies we have

$$< f \circ H, h > [A] = \overline{< f \circ H, h > [{^{\circ}A}]} \subseteq \overline{\Delta_{{^{\circ}A}}} = \Delta_A,$$

so that $f \circ H = h$. Similarly, $g \circ H = k$. Thus, $< f, g > \circ H = < h, k >$ and hence,

$$H = < f, g >^{-1} \circ < h, k > \in F(A).$$

Therefore, for arbitrary $h_1, \ldots, h_n \in F(A)$ the map $< h_1, \ldots, h_n >$ is a CL map $A \to A^n$ in the τ-topologies. Thus each $< h_1, \ldots, h_n > [A] \subseteq A^n$ is τ-clopen. Conversely, each τ-clopen set in A^n is of the form $\tau_n[\overline{Z}]$ for some $Z \subseteq {^{\circ}A}$ ($\overline{Z}$ = closure in A's original topology) and if $k \in F(A)$ satisfies $^{\circ}k[{^{\circ}A}] = Z$, then this τ-clopen is of the form $\tau_n \circ k[A]$ where $\tau_n \circ k$ can be written as $< h_1, \ldots, h_n >$ for certain $h_1, \ldots, h_n \in F(A)$. Thus, the τ-topologies given for the A^n coincide with the canonical topologies. This proves (a). Part (b) now follows from the fact that τ_1 and τ_2 are homeomorphisms both from original to canonical topologies and from original to original topologies.

In general, to show now that a given map $H : A^n \to A^m$ is CL it suffices now to show $H \circ \tau_n : A \to A^m$ is of the form $< h_1, \ldots, h_m >$ for some $h_1, \ldots, h_m \in F(A)$. In this manner, all projection maps are seen to be CL. The graph of any projection map or identity map is seen to be of the form $< h_1, \ldots, h_n > [A]$, for appropriate $h_1, \ldots, h_n \in F(A)$, and thus such a map is an element of the appropriate $F_{n,m}(A)$. This verifies the latter part of (c). Suppose $H \in F_{n,m}(A)$. Then $G(H) \subseteq A^{n+m}$ is clopen, hence, of the form

$$< h_1, \ldots, h_{n+m} > [A]$$

for certain $h_1, \ldots, h_{n+m} \in F(A)$. Since

$$H \circ < h_1, \ldots, h_n > = < h_{n+1}, \ldots, h_{n+m} > : A \to A^m,$$

where $< h_1, \ldots, h_n >$ is both onto and CL, and since $< h_{n+1}, \ldots, h_{n+m} >$ is CL, it follows that $H : A^n \to A^m$ is CL. This justifies the rest of (c).

Finally, let $\circ H : \circ A^n \to \circ A^m$ be an arbitrary map. Let the composition

$$\circ h = (\circ \tau_m^{-1}) \circ (\circ H) \circ (\circ \tau_n) : \circ A \to \circ A$$

have extension h in $F(A)$. Then

$$G(\circ H) = (\circ \tau_n \times \circ \tau_m)[G(\circ h)]$$

and hence

$$\overline{G(\circ H)} = (\tau_n \times \tau_m)[G(h)]$$

is the graph of a function $H : A^n \to A^m$ which is necessarily the unique extension of $\circ H$ in $F_{n,m}(A)$. This justifies (d).

The remaining assertions of the Lemma 6.4 follow from the fact that the graphs $G(f \circ g)$, $G(< f, g >)$, and $G(f \times g)$ are each expressible in terms of graphs $G(f)$ and $G(g)$ by forming intersections and direct and inverse images of projection maps. $\square$

Our last lemma narrows the field and prepares us for the final proof of Theorem 6.1.

LEMMA 6.5. *Suppose* $\circ A \subseteq A$ *are nonempty sets and that each finitary power* A^n *is given a* CL *topology with core* $(\circ A)^n$ *so that* $\Delta_A \subseteq A^2$ *is clopen and each projection* $\pi : A^n \to A^m$ *is a* CL *map. Then these topologies are the canonical topologies of a unique topological local internal domain structure on* A.

PROOF. The case for finite $\circ A$ is trivial, so we assume an infinite $\circ A$. The question of uniqueness is obvious. For each $n, m \geq 1$, let $F_{n,m}(A)$ be the collection of all maps $f : A^n \to A^m$ whose graphs $G(f) \subseteq A^{n+m}$ are clopen.

We first argue that each map $\circ f : \circ A^n \to \circ A^m$ extends to a unique element $f : A^n \to A^m$ of $F_{n,m}(A)$. Let W_1, W_2 and W_3 be A^n, A^m and A^m repectively. Let $\pi_{ij} : W_1 \times W_2 \times W_3 \to W_i \times W_j$ and $\pi_e : W_1 \times W_2 \to W_e$ be the obvious projections for $1 \leq i < j \leq 3$ and $1 \leq e \leq 2$. Denote by $\circ \pi_{ij}$ and $\circ \pi_e$ the corresponding projections on the cores. Let $G = \overline{G(\circ f)}$. Then $G \subseteq W_1 \times W_2$ is clopen. We argue that G is the graph of a mapping $f : W_1 \to W_2$. Since

$$\pi_1[G] = \overline{\pi_1[G(\circ f)]} = \overline{\circ W}_1 = W_1,$$

it follows that for all $p \in W_1$, there exists $q \in W_2$ such that $< p, q > \in G$. Also, since

$$\pi_{12}^{-1}[G] \cap \pi_{13}^{-1}[G]$$
$$= \overline{\circ \pi_{12}^{-1}[\circ G]} \cap \overline{\circ \pi_{13}^{-1}[\circ G]}$$
$$\subseteq \overline{\circ \pi_{23}^{-1}[\circ \Delta_{W_2}]} = \pi_{23}^{-1}[\Delta_{W_2}]$$

it follows that for all $< p, q >, < p, q' > \in G$, one has $q = q'$. Thus, $G = G(f)$ for some map $f : A^n \to A^m$ which, by definition, lies in $F_{n,m}(A)$.

We next argue that any element $f \in F_{n,m}(A)$ as a map $A^n \to A^m$ is CL. Clearly $^\circ G(f) = G(^\circ f)$ for some *partial* map $^\circ f ; ^\circ A^n \to {}^\circ A^m$ and since

$$\overline{^\circ A^n} = A^n = \pi_1[G(f)] = \overline{\pi_1[G(^\circ f)]}$$

it follows that this $^\circ f$ is total. Therefore, $f[^\circ A^n] \subseteq {}^\circ A^m$. Also, for $Z \subseteq {}^\circ A^n$ one has (continuing previous notation) that

$$\begin{aligned}
f[\overline{Z}] &= \pi_2[(\overline{Z} \times A^m) \cap G(f)] \\
&= \pi_2\overline{[(Z \times {}^\circ A^m) \cap G(^\circ f)]} \\
&= \overline{\pi_2[(Z \times {}^\circ A^m) \cap G(^\circ f)]} = \overline{f[Z]},
\end{aligned}$$

so that by Proposition 5.4, f is a CL map.

Familiar arguments now show that if f and g are elements of appropriate $F_{n,m}(A)$'s then, whenever defined, so are $f \circ g$, $< f, g >$, and $f \times g$. From this it is routine to verify that the given topologies on A and $A \times A$ induce a topological local internal domain structure on A with core $^\circ A$. Since $^\circ A$ is infinite, we have for each $n \geq 1$ there exist $f_1, \dots, f_n \in F_{1,1}(A)$ such that the map $< f_1, \dots, f_n > : A \to A^n$ is a homeomorphism, both from original to canonical, and original to original topologies, and thus it is clear that the topologies given on the A^n are the canonical ones. $\square$

Finally:

PROOF. [*of Theorem 6.1*] Assume that $\mathbf{A} = (A, \underline{A})$ is a local internal domain. By Lemma 6.5, to show the $\mathbf{A}$-topologies induce a topological local internal domain structure on A we need only show, for each n, that the $\mathbf{A}$-basis relations in $\underline{A}_n$ are *exactly* the $\mathbf{A}$-clopens in A^n (recall that $\Delta_A \in \underline{A}_2$). But but this follows from Propositions 2.5b, 4.1b, and 5.2a and the fact that $\mathbf{A}$ is local.

Conversely, suppose we have a nonempty set A with topological local internal domain structure imposed on it. For each n, let $\underline{A}_n$ be the collection of all subsets $R \subseteq A^n$ which are clopen in the canonical topology. Letting $\underline{A} = (\underline{A}_1, \underline{A}_2, \dots, \underline{A}_n, \dots)$ it is routine to verify that $\mathbf{A} = (A, \underline{A})$ is a local internal domain, and trivially, that the $\mathbf{A}$-topologies are the A-canonical topologies. The uniqueness of $\mathbf{A}$ follows from Lemma 6.5. $\square$

We end this chapter by considering how proper point set morphisms between local internal domains are also topologically determined.

THEOREM 6.2. *Let* $\mathbf{A}, \mathbf{A}' = (A, \underline{A}), (A', \underline{A}')$ *be local internal domains and suppose* $\mathbf{f} = (f, \underline{f}) : \mathbf{A} \to \mathbf{A}'$ *is a point set morphism between them which is proper. Then the set map* $f : A \to A'$ *satisfies the following properties:*

1) f *is* $(1 - 1)$
2) *both* $f : A \to A'$ *and* $f \times f : A^2 \to A'^2$ *are* CL_0 *maps with respect to the* $\mathbf{A}$- *and* $\mathbf{A}'$-*topologies.*

Conversely, let $f : A \to A'$ be any set map satisfying the foregoing properties. Then there exists a unique $\underline{f}$ such that $\mathbf{f} = (f, \underline{f})$ is a proper point set morphism $\mathbf{A} \to \mathbf{A}'$ of the local internal domains.

PROOF. Assume we are given a proper point set morphism $\mathbf{f} = (f, \underline{f}) : \mathbf{A} \to \mathbf{A}'$ between local internal domains. By proposition 2.5a, we know that each f_n is one to one and continuous. Also, by the transfer principle for point set structures (Theorem 2.1), f "preserves individuals", i.e., $f[{}^\circ A] \subseteq {}^\circ A'$ and hence, each f_n is a CL_0 map.

Conversely, suppppose we are given topological local internal domains A and A', and a mapping $f : A \to A'$ satisfying the properties (1) and (2). First we note that if we are given maps $h \in F(A)$ and $h' \in F(A')$ for which $({}^\circ f) \circ ({}^\circ h) = ({}^\circ h') \circ ({}^\circ f)$ is the case, then in fact $f \circ h = h' \circ f$ holds as well. Indeed, from $({}^\circ f) \circ ({}^\circ h) = ({}^\circ h') \circ ({}^\circ f)$ and f's injectivity we get

$$({}^\circ f \times {}^\circ f)^{-1}[G({}^\circ h')] = G({}^\circ h),$$

hence, from Proposition 5.3a, that

$$(f \times f)^{-1}[G(h')] = G(h) \text{ or}$$
$$(f \times f)[G(h)] \subseteq G(h'),$$

so for all $p \in A$, we have

$$< f(p), f(h(p)) > \ = \ < f(p), h'(f(p)) >$$

or, in other words, that $f \circ h = h' \circ f$ holds.

We use this observation to show that each of the induced maps $f_n : A^n \to A'^n$ is CL_0. Since the case where A is finite is trivial, we can assume that A is infinite. Then we can pick maps $h'_1, h'_2, \ldots, h'_n \in F(A')$ such that if $H' = < h'_1, h'_2, \ldots, h'_n > \in F_{1,n}(A')$ then the restriction of ${}^\circ H'$ to $f[{}^\circ A]$ is a bijection $f[{}^\circ A] \to f_n[{}^\circ A^n]$. Next, we can pick $h_1, h_2, \ldots, h_n \in F(A)$ such that in each case $({}^\circ f) \circ ({}^\circ h_i) = ({}^\circ h'_i) \circ ({}^\circ f)$. This means that $H = < h_1, h_2, \ldots, h_n > \in F_{1,n}(A)$ is a homeomorphism $A \to A^n$ for which $f_n \circ H = H' \circ f$ holds. This further means that $f_n = H' \circ f \circ H^{-1}$ is a composition of CL_0 maps and hence is CL_0 as well.

Letting $\mathbf{A}$, $\mathbf{A}' = (A, \underline{A})$, $(A', \underline{A}')$ be the associated local internal domains, we now construct from f a uniquely associated point set morphism $\mathbf{f} = (f, \underline{f}) : \mathbf{A} \to \mathbf{A}'$. For each $\mathbf{A}$-basis relation $R \in \underline{A}_n$ (i.e., any clopen $R \subseteq A^n$) define $\underline{f}_n(R) = \overline{f_n[R]}$. Since it is easy to check that $\overline{f_n[R]} = \overline{f_n[{}^\circ R]}$, we see that the map $\underline{f}_n : \underline{A}_n \to \underline{A}'_n$ preserves the appropriate boolean operations. It is routine to verify that the $\underline{f}_n$'s collectively preserve cartesian products and direct images by finitary projection and permutation maps. Thus, if we let $\underline{f} = (\underline{f}_1, \underline{f}_2, \cdots)$ and we set $\mathbf{f} = (f, \underline{f})$, we arrive at a canonically defined point set morphism $\mathbf{A} \to \mathbf{A}'$ of the local internal domains.

We need to argue that $\mathbf{f}$ is proper. Let $R' \in \underline{A}'_n$ satisfy $R' \subseteq \underline{f}_n(A^n)$. Then $R = f_n^{-1}[R']$ is an $\mathbf{A}$-clopen and lies in $\underline{A}_n$. But

$$^\circ(\underline{f}_n(R)) = {}^\circ(\overline{f_n[{}^\circ R]}) = f_n[{}^\circ R] = {}^\circ R',$$

and hence,

$$R' = \overline{{}^\circ R'} = \overline{{}^\circ(\underline{f}_n(R))} = \underline{f}_n(R),$$

which means that R' belongs to $\mathbf{f}[\mathbf{A}] \subseteq \mathbf{A}'$. Thus $\mathbf{f}[\mathbf{A}]$ is full in $\mathbf{A}'$ and $\mathbf{f}$ is proper.

By Proposition 2.5, any proper morphism $\mathbf{A} \to \mathbf{A}'$ is completely determined by its $A \to A'$, so that the $\mathbf{f}$ we have constructed is the unique one possible. $\square$

CHAPTER 7

Topological Determinacy of Internal Domains

The internal part ${}^*\mathbf{X} = ({}^*X, {}^*\underline{X})$ of a Robinson enlargement $* : \mathbf{X} \to \mathbf{X}'$ gives a typical example of an internal domain which is *nonlocal*. This shows up topologically. Although the ${}^*\mathbf{X}$-topologies turn both ${}^*X$ and ${}^*X^2$ into CL spaces, the projection maps ${}^*X \times {}^*X \to {}^*X$ are merely CL_0 and not CL maps. Instead, ${}^*X$ is the increasing union of ${}^*\mathbf{X}$-basis clopen compact subsets ${}^*X_n \, (n \geq 0)$ where the ${}^*\mathbf{X}$-induced topologies on ${}^*X_n$ and ${}^*X_n \times {}^*X_n$ turn each ${}^*X_n$ into a topological local internal domain in the sense of Chapter 6. In this manner, ${}^*\mathbf{X}$ is topologically a directed union of local internal domains.

This provides the key for discovering the topological determination of internal domains in general. The next theorem spells out the details.

THEOREM 7.1. *Let $\mathbf{A} = (A, \underline{A})$ be an internal domain. Then the $\mathbf{A}$-topologies on A and A^2 satisfy the following properties:*

1) *A and A^2 are CL spaces*
2) *there exists a collection $\{A_i : i \in \Gamma\}$ of subsets of A directed by inclusion which cover A such that for each $i \in \Gamma$, both $A_i \subseteq A$ and $A_i^2 \subseteq A^2$ are clopen subsets whose subspace topologies make A_i into a topological local internal domain.*

Conversely, if A is any nonempty set and topologies are given on A and A^2 which satisfy the foregoing properties, then there exists a unique $\underline{A}$ such that $\mathbf{A} = (A, \underline{A})$ is an internal domain whose $\mathbf{A}$-topologies on A and A^2 coincide with the ones given.

PROOF. Clearly an internal domain $\mathbf{A} = (A, \underline{A})$ is a directed union of its local full substructures, and these correspond exactly to the $\mathbf{A}$-basis sets (elements of $\underline{A}_1$). Each of these is seen to be a local internal domain. Since each $A'^n \subseteq A^n$ for $A' \in \underline{A}_1$ is $\mathbf{A}$-clopen, it follows from Proposition 5.1, that every A^n is a CL space. Thus (1) is satisfied. Also, if we take $\underline{A}_1$ as our collection $\{A_i : i \in \Gamma\}$ of subsets in A, it is clear that condition (2) is satisfied as well.

The nontrivial part of this theorem is the converse. I suppose now that we are given a nonempty set A with specified CL topologies on A and A^2. We shall call a collection $\mathcal{A} = \{A_i : i \in \Gamma\}$ of subsets in A satisfying condition (2) a *local internal domain covering* of A. We say that two local internal domain coverings $\mathcal{A}$, $\mathcal{A}'$ of A are *compatible* if there is a third $\mathcal{A}''$ such that $\mathcal{A}$, $\mathcal{A}' \subseteq \mathcal{A}''$. It is clear that to each local internal domain covering $\mathcal{A}$ we may canonically associate an internal pre-domain $\mathcal{A}(A) = (A, \underline{\mathcal{A}}(A))$ where for each n, $\underline{\mathcal{A}}_n(A)$ consists of all $R \subseteq A^n$ for which there exists $A' \in \mathcal{A}$ such that $R \subseteq A'^n$ is a clopen in the canonical A'-topology. Clearly, the $\mathcal{A}(A)$-topologies on A and A^2 coincide with the originals.

The proof of the converse of our theorem will amount to showing that *all* local internal domain coverings $\mathcal{A}$, $\mathcal{A}'$ of A are compatible. Thus, there can exist at most one *maximal* local internal domain covering of A, and this will obviously induce the unique internal domain structure $\mathbf{A} = (A, \underline{A})$ which is compatible with the given topologies on A and A^2. The axiom of choice guarantees the existence of the maximal cover as soon as any cover is known to be present.

I now further suppose that we are given a local internal domain covering $\mathcal{A}$. Our attention focuses on a particular clopen subset $\widehat{A} \subseteq A$ for which the subspace topologies on $\widehat{A}$ and $\widehat{A}^2$ determine $\widehat{A}$ as a topological local internal domain. The existence of the covering $\mathcal{A}$ already forces the projection maps $A \times A \rightarrow A$ to be continuous so that $\widehat{A}^2 \subseteq A^2$ is also clopen. I shall call such a subset $\widehat{A}$ *potential* . The idea is that $\widehat{A}$ might potentially be an element of an extended local internal domain covering $\mathcal{A} \subseteq \mathcal{A}'$. For this to happen, it is necessary and sufficient that $\widehat{A} \cup A'$ also be potential, for each $A' \in \mathcal{A}$. In this case, we say that the potential set $\widehat{A}$ is *compatible* with the cover $\mathcal{A}$.

I shall present a criterion for a potential set $\widehat{A}$ to be compatible with a cover $\mathcal{A}$ which is *independent* of the particular cover. Thus, $\widehat{A}$ will be compatible with some cover $\mathcal{A}$ if and only if it is compatible with *all* covers $\mathcal{A}$. From this it follows that any two coverings $\mathcal{A}$, $\mathcal{A}'$ of A are mutually compatible, and the theorem is proved.

The criterion for compatibility is based on the following technical concept.

DEFINITION 7.1. *Let X be any CL space. The* rank *of a point $p \in X$, written* $rank(p)$, *is the minimum of the cardinalities of subsets $Z \subseteq {}^\circ X$ such that $p \in \overline{Z}$.*

Of course, in any CL space X the set of subsets $Z \subseteq {}^\circ X$ whose closure $\overline{Z}$ contains a given point p forms an ultrafilter[1]. The simplest ultrafilters (called *principal* ultrafilters) are those containing a singleton. In this case the singleton would be $\{p\}$ and we would have $p \in {}^\circ X$. Points $p \in X - {}^\circ X$ will have more complicated ultrafilters. None of them contain a finite set. Some may not even contain countable sets, or sets of even higher cardinality. In this sense, the rank of a point $p \in X$ measures how complicated its ultrafilter is.

The motivation behind our criterion for compatibility is as follows: Without loss of generality, we can assume the covering $\mathcal{A}$ is maximal. Essentially, for po-

---

[1] An *ultrafilter* on a set Y is any collection $\mathcal{U}$ of its subsets which satisfies the finite intersection property and for any subset Y' in Y, contains either Y' or its complement $Y - Y'$. For more information concerning this concept consult [**5**].

tential set $\widehat{A}$ to be compatible with cover $\mathcal{A}$, it should contain points of arbitrary complexity (high rank) which the size (cardinality) of $^\circ\widehat{A}$ permits. Clearly, for each $\hat{p} \in \widehat{A}$ we have $\operatorname{rank}(\hat{p}) \leq \operatorname{card}(^\circ\widehat{A})$. On the other hand, if we are given a general point $p \in A$ for which $\operatorname{rank}(p) \leq \operatorname{card}(^\circ\widehat{A})$, then $\widehat{A}$ should contain a similar point with matching rank. This is because for some $A' \in \mathcal{A}$ with $\operatorname{card}(^\circ A') = \operatorname{rank}(\hat{p})$, we will have $p \in A'$, and compatibility with $\mathcal{A}$ will require that $\widehat{A} \cup A'$ is also potential, which would give us a $(1-1)$ CL map $f : A' \to \widehat{A}$ and hence, a point $\hat{p} = f(p) \in \widehat{A}$ of the same rank as p. For this reason, we define a potential set $\widehat{A} \subseteq A$ to have *rank compatibility* whenever

for all $p \in A$, if $\operatorname{rank}(p) \leq \operatorname{card}(^\circ\widehat{A})$, then there
exists $\hat{p} \in \widehat{A}$ such that $\operatorname{rank}(\hat{p}) = \operatorname{rank}(p)$.

We finish the proof of Theorem 7.1 by showing that a potential set $\widehat{A} \subseteq A$ with rank compatibility is actually compatible with any local internal domain covering $\mathcal{A}$. From this it quickly follows that any two local internal domain coverings of A are compatible.

We assume that our potential set $\widehat{A} \subseteq A$ has rank compatibility and that the covering $\mathcal{A}$ is maximal. We give each finitary power A^n its $\mathcal{A}(A)$-topology. To show that $\widehat{A}$ is compatible with $\mathcal{A}$, it suffices to pick an arbitrary element $A' \in \mathcal{A}$ and demonstrate that

$$\widetilde{A} = \widehat{A} \cup A' \subseteq A$$

is also a potential set.

I first argue for each $^\circ f : {}^\circ\widetilde{A} \to {}^\circ\widetilde{A}$, that $\overline{G(^\circ f)} \subseteq \widetilde{A}^2$ is the graph of a map $f : \widetilde{A} \to \widetilde{A}$. Suppose

$$< p, q >, < p, q' > \in \overline{G(^\circ f)}.$$

Let $\pi_{ij} : A^3 \to A^2$ for $1 \leq i < j \leq 3$ be the obvious projections. Being locally CL, they are CL_0 maps, and thus we have that

$$< p, q, q' > \in \pi_{12}^{-1}[\overline{G(^\circ f)}] \cap \pi_{13}^{-1}[\overline{G(^\circ f)}]$$
$$= \overline{{}^\circ\pi_{12}^{-1}[G(^\circ f)] \cap {}^\circ\pi_{13}^{-1}[G(^\circ f)]}$$
$$= \overline{{}^\circ\pi_{12}^{-1}[G(^\circ f)]} \cap \overline{{}^\circ\pi_{13}^{-1}[G(^\circ f)]}$$
$$\subseteq \overline{{}^\circ\pi_{23}^{-1}[\Delta_{{}^\circ\widetilde{A}}]} = \pi_{23}^{-1}[\overline{\Delta_{{}^\circ\widetilde{A}}}] = \pi_{23}^{-1}[\Delta_{\widetilde{A}}],$$

and hence that $q = q'$. Thus, $\overline{G(^\circ f)}$ is the graph of a *partial* map $f; \widetilde{A} \to \widetilde{A}$. We need to show that f in fact is defined on all of $\widetilde{A}$. Let $p \in \widetilde{A}$. Since $^\circ f^{-1}[^\circ\widehat{A}]$ and $^\circ f^{-1}[^\circ A']$ cover $^\circ\widetilde{A}$ and $\mathcal{A}$ is assumed to be a maximal local internal domain cover, we can pick $A_1 \in \mathcal{A}$ such that $p \in A_1 \subseteq \widetilde{A}$ and,

either $^\circ f[^\circ A_1] \subseteq {}^\circ\widehat{A}$ or $^\circ f[^\circ A_1] \subseteq {}^\circ A'$.

If the latter holds, then by $\mathcal{A}$'s maximality,

$$A'' = A_1 \cup A' \in \mathcal{A}$$

and there will exist $q \in A''$ for which

$$< p, q > \in \overline{G(^\circ f|_{^\circ A_1})} \subseteq \overline{G(^\circ f)}$$

showing that $f(p)$ is defined. If $^\circ f[^\circ A_1] \subseteq {^\circ \widehat{A}}$ and $p \in \widehat{A}$, then we can assume $A_1 \subseteq \widehat{A}$, and then $f(p)$ is defined for similar reasons.

Suppose $^\circ f[^\circ A_1] \subseteq {^\circ \widehat{A}}$ and that $p \in A' - \widehat{A}$. This is the more subtle case. We can assume that $A_1 \subseteq A'$. Let $^\circ f' = {^\circ f|_{^\circ A_1}}$. Then $^\circ f'$ can be expressed as a composition $^\circ f' = (^\circ h_2) \circ (^\circ h_1)$ where

$$^\circ h_1 : {^\circ A_1} \to {^\circ A_1}, \text{ and } {^\circ h_2} : {^\circ A_1} \to {^\circ \widehat{A}},$$

and $^\circ h_2$ is $(1-1)$ on $^\circ h_1[^\circ A_1]$. Pick $p' \in A_1$ such that

$$< p, p' > \in \overline{G(^\circ h_1)}.$$

Now suppose there exists $q \in \widehat{A}$ such that $< p', q > \in \overline{G(^\circ h_2)}$. Then, again letting $\pi_{ij} : A^3 \to A^2$ be the usual projections, we have that

$$< p, p', q > \in \pi_{12}^{-1}[\overline{G(^\circ h_1)}] \cap \pi_{23}^{-1}[\overline{G(^\circ h_2)}]$$
$$= {^\circ \pi_{12}^{-1}}[\overline{G(^\circ h_1)}] \cap {^\circ \pi_{23}^{-1}}[\overline{G(^\circ h_2)}]$$
$$= {^\circ \pi_{12}^{-1}}[\overline{G(^\circ h_1)}] \cap {^\circ \pi_{23}^{-1}}[\overline{G(^\circ h_2}]$$
$$\subseteq {^\circ \pi_{13}^{-1}}[\overline{G(^\circ f')}] = \pi_{13}^{-1}[\overline{G(^\circ f')}] \subseteq \pi_{13}^{-1}[\overline{G(^\circ f)}],$$

and hence $f(p)$ is defined.

Thus, I need to verify the existence of $q \in \widehat{A}$ such that $< p', q > \in \overline{G(^\circ h_2)}$. Since $p' \in {^\circ h_1[^\circ A_1]}$ and $^\circ h_2$ is $(1-1)$ on $^\circ h_1[^\circ A_1]$, it is clear that

$$rank(p') \le card(^\circ \widehat{A}).$$

By the rank compatibility assumed for $\widehat{A}$, there exists $q' \in \widehat{A}$ such that

$$rank(p') = rank(q').$$

Pick A_2, $A_3 \in \mathcal{A}$ such that $p' \in A_2 \subseteq {^\circ h_1[^\circ A_1]}$ and $q' \in A_3 \subseteq \widehat{A}$ and

$$card(A_2) = card(A_3) = rank(p').$$

Then we can write $^\circ h_2|_{A_2}$ as a composition $(^\circ h_4) \circ (^\circ h_3)$ where

$$^\circ h_3 : {^\circ A_2} \to {^\circ A_3} \text{ and } {^\circ h_4} : {^\circ A_3} \to \widehat{A}.$$

Picking $q' \in A_3$ and $q \in \widehat{A}$ such that

$$< p', q' > \in \overline{G(^\circ h_3)} \text{ and } < q', q > \in \overline{G(^\circ h_4)}$$

(since $A_3 \subseteq \widehat{A}$, q obviously exists), we can apply a previous argument to the tuple $< p', q', q >$ and conclude that

$$< p', q > \in \overline{G(^\circ h_2|_{A_2})} \subseteq \overline{G(^\circ h_2)}.$$

This shows that $f(p)$ is defined in all cases.

Thus each set map $^\circ f : {}^\circ\widetilde{A} \to {}^\circ\widetilde{A}$ extends to a unique map $f : \widetilde{A} \to \widetilde{A}$ whose graph is clopen.

From this, it will follow that the projections $\pi_i : \widetilde{A} \times \widetilde{A} \to \widetilde{A}$ $i = 1, 2$ are CL maps. Indeed, let $Z \subseteq {}^\circ\widetilde{A} \times {}^\circ\widetilde{A}$ and $p \in \overline{\pi_1[Z]}$. I shall show that $p \in \pi_1[\overline{Z}]$. Pick $^\circ f : {}^\circ\widetilde{A} \to {}^\circ\widetilde{A}$ such that

$$G(^\circ f|_{\pi_1[Z]}) \subseteq Z$$

and also pick $q \in \widetilde{A}$ such that $< p, q > \in \overline{G(^\circ f)}$. Then

$$< p, q > \in \overline{G(^\circ f)} \cap \overline{\pi_1[Z]} \times \widetilde{A} = \overline{G(^\circ f) \cap \pi_1[Z] \times {}^\circ\widetilde{A}}$$
$$= \overline{G(^\circ f) \cap \pi_1[Z] \times {}^\circ\widetilde{A}} = \overline{G(^\circ f|_{\pi_1[\overline{Z}]})} \subseteq \overline{Z}$$

and hence, $p \in \pi_1[\overline{Z}]$. This shows for every $Z \subseteq {}^\circ\widetilde{A} \times {}^\circ\widetilde{A}$ that $\overline{\pi_1[Z]} \subseteq \pi_1[\overline{Z}]$ and since π_1 is a CL_0 map it must, by Proposition 5.3b, be a CL map. Also, if $\sigma : \widetilde{A}^2 \to \widetilde{A}^2$ is the nontrivial permutation map, then it is a bijective locally CL and hence, CL_0 map which, by Proposition 5.3c, is CL. Thus $\pi_2 = \pi_1 \circ \sigma$ is the composition of CL maps, and is also CL.

Recall that our goal is to show that $\widetilde{A}$ is a potential set in A. For the remainder of our argument, we let $F(\widetilde{A})$ denote the collection of all maps $f : \widetilde{A} \to \widetilde{A}$ whose graphs are clopen.

It is trivial that $id_{\widetilde{A}} \in F(\widetilde{A})$. We need also to check that $\widetilde{A} \times \widetilde{A}$ is covered by sets of the form $< f, g > [\widetilde{A}]$ for $f, g \in F(\widetilde{A})$. Given $p, q \in \widetilde{A}$, pick $A'' \in \mathcal{A}$ (appealing to $\mathcal{A}$'s maximality) such that

$$p, q \in A'' \subseteq \widetilde{A}.$$

Then we can pick $f, g \in F(\widetilde{A})$ such that

$$f[A''], \ g[A''] \subseteq A''$$

and there exists $r \in A''$ such that $< r, p, q >$ lies in

$$\overline{G(< {}^\circ f|_{\circ A''}, {}^\circ g|_{\circ A''} >)} = G(< f|_{A''}, g|_{A''} >).$$

Thus, $< f, g > (r) = < p, q >$, and $\widetilde{A} \times \widetilde{A}$ is seen to be covered by sets of the form $< f, g > [\widetilde{A}]$, for $f, g \in F(\widetilde{A})$.

Finally, I need to show, for any $f, g \in F(\widetilde{A})$, that $f \times g : \widetilde{A}^2 \to \widetilde{A}^2$ is CL. Clearly, $f \times g$ — being locally CL— is at least CL_0. Let $Z \subseteq {}^\circ\widetilde{A}^2$ and pick $< p, q > \in \overline{(f \times g)[Z]}$. By Proposition 5.3b, we need to show $< p, q > \in (f \times g)[\overline{Z}]$. Without loss of generality, we can assume that $f \times g|_Z$ is one to one. Pick $r \in \widetilde{A}$ and $h, k \in F(\widetilde{A})$ such that

$$< h, k > (r) = < p, q > \ and < h, k > [\widetilde{A}] \subseteq \overline{(f \times g)[Z]}.$$

Pick $h', k' \in F(\widetilde{A})$ such that

$$< {}^\circ h', {}^\circ k' > = ({}^\circ f \times {}^\circ g)|_Z^{-1} \circ < {}^\circ h, {}^\circ k > .$$

Let $< h', k' > (r) = < p', q' >$. Then, letting $\pi_{ij} : \widetilde{A}^5 \to \widetilde{A}^2$ for $1 \leq i < j \leq 5$ be the obvious projections (which are CL_0), we have that the 5-tuple

$$< r, p', q', p, q > \in \widetilde{A}^5$$

is an element of the set

$$
\begin{aligned}
&\pi_{12}^{-1}[G(h')] \cap \pi_{13}^{-1}[G(k')] \cap \pi_{14}^{-1}[G(h)] \cap \pi_{15}^{-1}[G(k)] \\
&= \pi_{12}^{-1}\overline{[G(^\circ h')]} \cap \pi_{13}^{-1}\overline{[G(^\circ k')]} \cap \pi_{14}^{-1}\overline{[G(^\circ h)]} \cap \pi_{15}^{-1}\overline{[G(^\circ k)]} \\
&= \overline{{}^\circ\pi_{12}^{-1}[G(^\circ h')] \cap {}^\circ\pi_{13}^{-1}[G(^\circ k')] \cap {}^\circ\pi_{14}^{-1}[G(^\circ h)] \cap {}^\circ\pi_{15}^{-1}[G(^\circ k]} \\
&= {}^\circ\pi_{12}^{-1}[G(^\circ h')] \cap {}^\circ\pi_{13}^{-1}[G(^\circ k')] \cap {}^\circ\pi_{14}^{-1}[G(^\circ h)] \cap {}^\circ\pi_{15}^{-1}[G(^\circ k)] \\
&\subseteq {}^\circ\pi_{24}^{-1}[G(^\circ f)] \cap {}^\circ\pi_{35}^{-1}[G(^\circ g)] \cap {}^\circ\pi_{23}^{-1}[Z] \\
&= \pi_{24}^{-1}[G(f)] \cap \pi_{35}^{-1}[G(g)] \cap \pi_{23}^{-1}[\overline{Z}].
\end{aligned}
$$

Hence $< p', q' > \in \overline{Z}$, $f(p') = p$, $g(q') = q$ and thus, $< p, q > \in (f \times g)[\overline{Z}]$. Therefore, $f \times g : \widetilde{A}^2 \to \widetilde{A}^2$ is a CL map.

This concludes the proof that $\widetilde{A} = \widehat{A} \cup A' \subseteq A$ is potential, and thus, that rank compatibility implies arbitrary compatibility. $\square$

Using Theorem 7.1, we can now offer a purely topological definition of internal domains. We shall consider it the primary definition of this concept during the remainder of this work.

DEFINITION 7.2. *An* internal domain *is any set A with topologies specified on A and A^2 which satisfy the properties 1) and 2) of Theorem 7.1. If these topologies satisfy properties 1) through 5) of Theorem 6.1 (i.e., the collection $\{A_i : i \in \Gamma\}$ in Theorem 7.1 can be taken to be $\{A\}$), then A is called a* local internal domain. *In either case, the associated point set structure $\mathbf{A} = (A, \underline{A})$, which is an internal domain in the original sense, is called the* canonical structure. *The* core *of the canonical structure $\mathbf{A}$ is $^\circ\mathbf{A} = (^\circ A, {}^\circ\underline{A})$, where for all n, $^\circ\underline{A}_n = \{^\circ R \subseteq {}^\circ A^n : R \in \underline{A}_n\}$. The obvious (proper dominant) point set morphism $* : {}^\circ\mathbf{A} \to \mathbf{A}$ is the* standard copy correspondence *. For every index set I, the* canonical topology *on A^I is its $\mathbf{A}$-topology. A set or relation $R \subseteq A^n$ is said to be* local *if $R \subseteq A'^n$ for some $\mathbf{A}$-basis set $A' \in \underline{A}_1$. A partial or total mapping $f; A^n \to A^m$ is* local preserving *if $f[R] \subseteq A^m$ is local whenever $R \subseteq A^n$ is local and contained in f's domain. For each n, $m \geq 1$, we write $F_{n,m}(A)$ for $\mathbf{A}$'s collection $F_{n,m}(\mathbf{A})$ of local maps $A^n \to A^m$. The internal domain A is κ-saturated *or* locally κ-saturated *if its canonical structure $\mathbf{A} = (A, \underline{A})$ is.*

The following is relevant topological information concerning the elements of $F_{n,m}(A)$:

PROPOSITION 7.1. *Let A be an internal domain. Then for each n, $m \geq 1$, the elements of $F_{n,m}(A)$ are exactly the maps $f : A^n \to A^m$ whose graphs $G(f) \subseteq A^{n+m}$ are clopen. These are local preserving continuous maps whose restriction to any local clopen $R \subseteq A^n$ is CL. Furthermore, any local preserving map $^\circ f : {}^\circ A^n \to {}^\circ A^m$ extends to a unique map $f : A^n \to A^n$ lying in $F_{n,m}(A)$.*

PROOF. The only nonroutine item needing proof here is the fact that any map $f : A^n \rightarrow A^m$ with clopen graph $G(f) \subseteq A^{n+m}$ is necessarily a local preserving map. Such maps clearly include all identity and projection maps, and have the usual closure properties associated with the function families $F_{n,m}(A)$. Routine considerations also show they are locally CL maps and hence are CL_0. Thus, the argument for local preservation readily reduces to the case where $n = m = 1$, which we now assume. Let $A' \subseteq A$ be an arbitrary local clopen. It will suffice for us to show that $f[A'] \subseteq A$ is also local. Without loss of generality, we can assume that $°f$ is one to one on $°A'$.

We first argue that $f[A'] \subseteq A$ is clopen. To show this, it suffices to show that $f[A'] \cap A''$ is clopen for every choice of local clopen $A'' \subseteq A$. In fact, it will suffice to show this for the case when $A' \subseteq A''$. But then

$$f[A'] \cap A'' = \pi_2[G(f) \cap A' \times A''],$$

which is clopen.

Next we show that f itself is one to one on A'. Suppose p, $p' \in A'$ are such that $f(p) = f(p') = q$. Pick local clopen $A'' \subseteq A$ which contains A' and the element q. Since f is one to one on $°A' \subseteq °A''$ and p, $p' \in A'$, it is clear that p and p' must be identical and thus on A', f is $(1 - 1)$.

At this point we have that $f|'_A$ is a bijective CL_0 map onto clopen $f[A']$ and hence, by Proposition 5.3c, is a CL isomorphism. Similarly, $f \times f$ induces a CL isomorphism from $A' \times A'$ to clopen $f[A'] \times f[A']$. Both projections

$$\pi_i : A^2 \rightarrow A \quad i = 1, 2$$

are now readily seen to be CL maps on $f[A'] \times f[A']$, and thus $f[A'] \subseteq A$ is a local internal domain with respect to the A-and $A \times A$-subspace topologies.

Thus, in the sense of the proof of Theorem 7.1, $f[A'] \subseteq A$ is a "potential set". Since $f[A']$ is isomorphic to A' and A' is a local clopen, $f[A']$ must also satisfy rank compatibility and hence, is itself a local clopen. Thus, f is seen to preserve localness. $\square$

Since the standard copy correspondence is a dominant morphism of point set structures, a *transfer principle* for internal domains is automatic. This formulates as follows: If we let A be an internal domain, then its language **L** shall be that associated with its canonical structure $\mathbf{A} = (A, \underline{A})$. Recall the language **L** has elements $a \in °A$ for individual constant symbols, elements $R \in \underline{A}_n$ for n-ary predicate symbols, and elements $f \in F_{n,1}(\mathbf{A})$ $(= F_{n,1}(A))$ for n-ary function symbols. The "core" language $°\mathbf{L}$ $(=$ the language of $°\mathbf{A} = (°A, °\underline{A}))$ has the corresponding elements a, $°R$ and $°f$ for constant, predicate and function symbols. A *local* formula of **L** shall be one whose quantifiers are local, i.e., of the form $(\forall x \in A')$ or $(\exists x \in A')$ for $A' \in \underline{A}_1$. A *local* formula of $°\mathbf{L}$ shall be one whose quantifiers have the corresponding form $(\forall x \in °A')$ or $(\exists x \in °A')$. There is the natural $*$-interpretation of $°\mathbf{L}$ in **L**: to each local formula φ of $°\mathbf{L}$

corresponds the local formula $^*\varphi$ of $\mathbf{L}$ obtained by replacing each symbol $^\circ R$ or $^\circ f$ in φ by the symbol R or f. One has the obvious interpretation of local sentences from $^\circ\mathbf{L}$ as statements about $^\circ\mathbf{A} = (^\circ A, ^\circ\underline{A})$ and of local sentences from $\mathbf{L}$ as statements about $\mathbf{A} = (A, \underline{A})$.

The transfer principle for internal domains becomes:

THEOREM 7.2. [transfer principle for internal domains] *Let A be an internal domain and let φ be a local sentence in the language $^\circ\mathbf{L}$ whose interpretation in $\mathbf{L}$ is $^*\varphi$. Then φ is true iff $^*\varphi$ is true.*

PROOF.   □

An important element in the story of topological determinacy for internal domains remains, namely, their proper morphisms:

THEOREM 7.3. *Let A and A' be internal domains and let $\mathbf{f} = (f, \underline{f})$ be a point set morphism $\mathbf{A} \to \mathbf{A}'$ of their canonical structures which is proper. Then the set map $f : A \to A'$ satisfies the following properties:*
   1) *f is $(1-1)$ and preserves local sets*
   2) *$f : A \to A'$ and $f_2 : A^2 \to A'^2$ are both CL_0 maps.*
In which case all the maps $f_n : A^n \to A'^n$ are are $(1-1)$ and CL_0. Conversely, let $f : A \to A'$ be any set map satisfying properties 1) and 2). Then there exists a unique $\underline{f}$ such that $\mathbf{f} = (f, \underline{f})$ is a proper point set morphism $\mathbf{A} \to \mathbf{A}'$ of their canonical structures. This $\mathbf{f}$ is determined by the property that for $R \in \underline{A}_n$, $\underline{f}_n(R) = \overline{f_n[R]} \subseteq A'^n$ ($\mathbf{A}'$-closures).

PROOF. The theorem readily follows from Theorem 6.2.   □

DEFINITION 7.3. *We call a mapping $f : A \to A'$ between two internal domains satisfying properties 1) and 2) of Theorem 7.3 a* morphism *of the internal domains. If f is a CL map we further call it a* CL-morphism*. If $A \subseteq A'$ and the inclusion is a morphism of internal domains, we call A an internal* subdomain *of A' and say A' is an* extension *of A.*

We finish this chapter with a few results which will be needed later in Part 3.

THEOREM 7.4. *Let $\mathbf{f} : \mathbf{A} \to \mathbf{A}'$ be a proper dominant point set morphism from an internal domain $\mathbf{A}$ to a point set structure $\mathbf{A}'$. Then $\mathbf{A}'$ is an internal domain.*

PROOF. Clearly, $\mathbf{A}'$ is already an internal *pre*-domain. Let $\mathbf{A}' \subseteq \mathbf{A}''$ be the unique topologically conservative extension of $\mathbf{A}'$ to an internal domain $\mathbf{A}''$. Writing $\mathbf{A} = (A, \underline{A})$, $\mathbf{A}' = (A', \underline{A}')$ and $\mathbf{A}'' = (A'', \underline{A}'')$, we can make the identifications $A \subseteq A' = A''$. Of course the $\mathbf{A}'$-topologies and the $\mathbf{A}''$-topologies on A' are identical, so that $^\circ A' = ^\circ A''$. Since $\mathbf{f}$ is proper and dominant, it follows easily that $^\circ A = ^\circ A'$. By Proposition 2.5, the $\mathbf{A}$-topologies are just the $\mathbf{A}'$-subspace topologies. To show that $\mathbf{A}'$ is an internal domain, we need to show that $\mathbf{A}' = \mathbf{A}''$. We know for every n, that $\underline{A}'_n \subseteq \underline{A}''_n$. It will suffice to show equality when $n = 1$. Let $\widehat{A}'' \in \underline{A}''_1$ be arbitrary and put $\widehat{A} = \widehat{A}'' \cap A$. Choose an $R \subseteq {}^\circ\widehat{A}^n$ and any projection map $\pi : A'^n \to A'^m$. Then in the $\mathbf{A}'$-topologies

$$\pi[\mathbf{A}\text{-closure of } R] = \pi[\overline{R} \cap \widehat{A}^n] =$$
$$\bigcup_{R' \subseteq R,\, R' \in {}^\circ\underline{A}_n} \pi[\overline{R}' \cap \widehat{A}^n] =$$
$$\bigcup_{R' \subseteq R,\, R' \in {}^\circ\underline{A}_n} \pi[\overline{R}'] \cap \widehat{A}^m =$$
$$\pi[\overline{R}] \cap \widehat{A}^m = \overline{\pi[R]} \cap \widehat{A}^m =$$
$$\mathbf{A}\text{-closure of } \pi[R],$$

and since $\Delta_{\widehat{A}}$ is $\mathbf{A}$-clopen, it follows from Lemma 6.5, that in the $\mathbf{A}$-topologies $\widehat{A}$ is a local internal domain. Thus, in the sense of the proof of Theorem 7.1, $\widehat{A}$ is "potential" in A. But for any $A_0 \in \underline{A}_1$, $\overline{A_0} \in \underline{A}'_1 \subseteq \underline{A}''_1$, and we have that $\overline{A_0} \cup \widehat{A}'' \in \underline{A}''_1$, so that $\overline{A_0} \cup \widehat{A} = (\overline{A_0} \cup \widehat{A}'') \cap A$ is also "potential", and so indeed, $\widehat{A}$ is "compatible". Since $\mathbf{A}$ is an internal domain, it follows that $\widehat{A} \in \underline{A}_1$ and hence, $\widehat{A}'' = \overline{\widehat{A}} \in \underline{A}'_1$. Thus, $\underline{A}'_1 = \underline{A}''_1$ and hence, $\mathbf{A}' = \mathbf{A}''$ is an internal domain as a point set structure. $\square$

THEOREM 7.5. *Let A be an internal domain and κ an infinite cardinal number. Then a morphism $f : A \to A'$ into a κ-saturated internal domain exists.*

PROOF. Use Theorem 2.2 to get a proper dominant point set morphism $\mathbf{f} = (f, \underline{f}) : \mathbf{A} \to \mathbf{A}'$ where $\mathbf{A} = (A, \underline{A})$ is A's canonical structure and $\mathbf{A}' = (A', \underline{A}')$ is a κ-saturated point set structure. By Theorem 7.4, $\mathbf{A}'$ is an internal domain. $\square$

PROPOSITION 7.2. *For any given internal domain A and set ${}^\circ A'$ for which $A \cap {}^\circ A' = {}^\circ A$ there exists extension $A \subseteq A'$ to a local internal domain which has ${}^\circ A'$ as core and A as open subset.*

PROOF. Assume the set ${}^\circ A'$ satisfies $A \cap {}^\circ A' = {}^\circ A$. Let $\widetilde{A}'$ consist of all ordered triples $({}^\circ f, {}^\circ B, p)$ where ${}^\circ B$ is the core of a local clopen $B \subseteq A$ which contains p as an element and ${}^\circ f : {}^\circ B \to {}^\circ A'$ is an arbitrary map. Two such triples $({}^\circ f, {}^\circ B, p)$ and $({}^\circ f', {}^\circ B', p')$ are called *equivalent* if for some map $g \in F_{1,1}(A)$ and clopen set $B_1 \subseteq B$, all of

$$p \in B_1,\ {}^\circ g[{}^\circ B_1] \subseteq {}^\circ B',\ g(p) = p',$$
$${}^\circ g|_{{}^\circ B_1} \text{ is } (1-1), \text{ and } ({}^\circ f') \circ ({}^\circ g|_{{}^\circ B_1}) = {}^\circ f|_{{}^\circ B_1}$$

hold. One checks that this indeed gives an equivalence relation on $\widetilde{A}'$. We let $[{}^\circ f, {}^\circ B, p]$ denote the equivalence class of $({}^\circ f, {}^\circ B, p)$ and write A' for the set of all such equivalence classes. The assignment of $p' \in {}^\circ A'$ to any $[{}^\circ f, {}^\circ B, p]$ where ${}^\circ f$ is a constant function with single value p' is well defined and one to one, so that a canonical identification ${}^\circ A' \subseteq A'$ is made. Also, the assignment of $p \in A$ to any $[{}^\circ f, {}^\circ B, p]$ where ${}^\circ f$ is an identity map is also well defined, one to one and canonically identifies $A \subseteq A'$. One checks that the two identifications are compatible. In a similar manner, one gets for each $[{}^\circ f, {}^\circ B, p]$ a well defined map $f : B \to A'$ sending $p' \in B$ to $[{}^\circ f, {}^\circ B, p']$. This map is one to one whenever ${}^\circ f$ is.

Such a one to one map f can be used to impose a local internal domain structure on the set $f[B] \subseteq A'$. These local internal domains are checked to be compatible on overlaps and one routinely finds that an internal pre-domain structure $\mathbf{A}'$ on A' having core $^\circ A'$ is canonically induced. One checks that the $\mathbf{A}'$-topologies on the A'^n's satisfy the hypotheses of Lemma 6.5, so in fact, A' is a *local* internal domain. Clearly, A is identified as an open internal subdomain of A'. $\square$

PROPOSITION 7.3. *The class of internal domains is closed under direct limits. These limits preserve local compactness and local ω-saturatedness. If the morphisms are dominant, then the limit will preserve localness, compactness and ω-saturation.*

PROOF. Let $(\Gamma, <)$ be a directed set and for each $i \in \Gamma$, let A_i be a internal domain so that for $i, j \in \Gamma$, with $i < j$, one has a morphism $A_i \to A_j$. Assume that these morphisms are compatible in the sense of the discussion of direct limits in Section 2.3. Using Theorem 7.3, we get induced proper point set morphisms $\mathbf{A}_i \to \mathbf{A}_j$ of the canonical structures. These morphisms also satisfy compatibility. Let $\mathbf{A} = (A, \underline{A})$ be the direct limit of these point set structures. It is easily seen that $\mathbf{A}$ is an internal pre-domain. The $\mathbf{A}$-topologies induced on A and A^2 thus give an internal domain structure on A. The canonical point set morphisms $\mathbf{A}_i \to \mathbf{A}$ are all proper and induce appropriate morphisms $A_i \to A$ of the internal domains. Theorem 7.3 implies that these morphisms make A into a direct limit of the A_i's.

To finish the proof, it will suffice to assume all the A_i's are local, that all morphisms $A_i \to A_j$ are dominant and show that localness, compactness, and ω-saturation are each preserved in the limit. We can assume $A_i \subseteq A_j \subseteq A$ for all $i < j$ and, by Proposition 5.3d, that all subtopologies are seen to be subspace topologies. The induced inclusions $A_i^I \subseteq Aj^I \subseteq A^I$, for any index set I, are also seen to share this property. Since morphisms are dominant we have $^\circ A_i = {}^\circ A_j = {}^\circ A$ for all $i < j$ and hence, A is local. If any of the A_i's are compact then A has a dense compact subset and since A is zero dimensional, it easily follows that A itself is compact. Similarly, if any A_i^I is compact, then so is A^I.

We now assume that each A_i is ω-saturated. Thus A^ω is compact. We need to show that any given projection $\pi : A^\omega \to A^I$, with $I \subseteq \omega$ finite, preserves closures of points. In fact, let $Z \subseteq A^\omega$ satisfy $Z \subseteq A_i^\omega$ for some i (any singleton $\{f\} \subseteq A^\omega$ does this). Without loss of generality, we can assume $Z \subseteq A_i^\omega$ holds for all i. Then, recalling from Proposition 2.3 that π's restriction to any A_i^ω is a continuous closed mapping, we have

$$\pi[\overline{Z}] = \pi[\bigcup_{i \in \Gamma} \overline{Z} \cap A_i^\omega]$$
$$= \bigcup_{i \in \Gamma} \pi[\overline{Z} \cap A_i^\omega]$$
$$= \bigcup_{i \in \Gamma} \overline{\pi[Z] \cap A_i^I} = \overline{\pi[Z]},$$

so that π preserves closures of points and hence, A is ω-saturated. $\square$

Part 3

Set Theoretic Aspects

CHAPTER 8

Introduction

Historically, Robinson's enlargements (see [**20**]) provided the first and most durable vehicle for "non"standard mathematics. Nevertheless, disadvantages were noted. As they required some familiarity with mathematical logic or model theory, these enlargements remained inaccessible to most working mathematicians. Moreover, different enlargements were needed for different applications. One was never enough. Enlargements were also restricted to a fragment of set theory. By the mid 1970's, the idea of a global vehicle for "non"standard mathematical practice had emerged. The need was felt for a single axiomatic "non"standard set theory which would formalize "non"standard mathematics in much the same way that conventional axiomatic set theory has been used to formalize standard mathematics. In [**18**] Nelson introduced Internal Set Theory (IST for short). In [**11**] Hrbáček proposed two "non"standard set theories NS_1 and NS_2. In a series of papers culminating in [**14**] Kawai developed a "non"standard set theory NST. More recently, in [**9**], Fletcher has proposed a "stratified" "non"standard set theory $SNST$.

It is a curious and stubborn fact that none of these proposals succeeded. Indeed, except in the case of Nelson's IST, there is no evidence that any of these proposed vehicles were ever even used. Nevertheless, the original idea persists and these earlier attempts provide ongoing insight concerning how the "correct" vehicle for global "non"standard mathematical practice might look.

In Part 3, I shall begin by reviewing this state of affairs in some detail. In Chapter 9, I recall for the reader the axiomatics of standard set theory (ZFC and its variants) and review the basic concepts and results from this theory which will be needed for later discussion. In Chapter 10, I give a unified description of the "non"standard set theories IST, NST, NS_1, and NS_2 (saving discussion of Fletcher's $SNST$ for later). In Chapter 11, I offer new proofs that each of these theories (IST, NST, NS_1, NS_2) is a conservative extension of ZFC. This means that any theorem provable in one of these extended theories which can be expressed as an ordinary statement about conventional sets is already a theorem provable in ZFC. The method of showing this involves working in a suitable conservative extension of ZFC, which I call AST (for **A**mbient **S**et

Theory). Using the topological tools developed in Part 2, I construct models of each "non"standard theory in the AST-universe and in a manner which uniformly guarantees the theory's conservativity over ZFC.

In Chapter 12, we pause to take a philosophical view of the overall project of finding a suitable global vehicle for "non"standard mathematical practice. I review the strengths and weaknesses of each of IST, NST, NS_1, and NS_2 in light of this goal. A key feature in the discussion is two broad limitations discovered by Hrbáček in [**11**] which apply to "non"standard set theories generally. I discuss Fletcher's more recent attempt in [**9**] to improve on the previous theories while avoiding the Hrbáček limitations. I argue that his theory $SNST$ has its own problems, but nonetheless, contributes useful ideas. I eventually draw up a basic shopping list of desired features which any "correct" global vehicle should have, should there be one. This is to include a certain flexibility of Robinson's original enlargements which is misssing in *all* the current "non"standard set theories[1]. I end by proposing a simple yet nontrivial modification of Fletcher's $SNST$ which remedies this defect and, in fact, combines all the features mentioned on the shopping list. This new set theory is named EST (for **E**nlargement **S**et **T**heory). Chapter 12 concludes with a description of EST's axiomatics.

Chapter 13 is devoted to a proof that EST is also conservative over ZFC. The approach is as before. Using the topological methods from Part 2, I construct a model of EST within the AST-universe in a manner that guarantees conservativity. Indeed, this time the method of model construction reveals numerous features which can be also conservatively added to EST, if so desired by a practitioner.

Discussion of EST's future usefulness to the greater mathematical community is saved for the concluding remarks in Chapter 14. The discussion is frank and open ended.

---

[1]This defect was pointed out to me by Hrbáček.

CHAPTER 9

Standard Set Theory

The "naive" set theory of most mathematicians finds a convenient codification in Zermelo Fraenkel Set Theory with the axiom of choice (ZFC for short). In this section we will review this set theory as well as a number of its variations. Further details can be found in [13]. I shall assume the concepts and notation from model theory which were discussed in Section 2.1. As we shall be discussing a number of related set theories, I shall keep our syntax and semantics informal and flexible.

A language **L** appropriate for a set theory will include a distinguished binary predicate $\in$ to indicate *membership*. An atomic formula $\in (s, t)$ is written $s \in t$ and reads "s is an element of t". The objects of the set theory are (at the very least) *classes*. We let $A, B, C, \ldots$ be informal variables which range over classes. Basic to all set theories is the *axiom of extensionality*, namely, that classes having the same elements are identical. In symbols:

$$(\forall C)[C \in A \leftrightarrow C \in B] \rightarrow A = B.$$

A class A is a *subclass* of a class B (written $A \subseteq B$) if each element of A is an element of B. Thus, by extensionality

$$A = B \text{ iff } A \subseteq B \text{ and } B \subseteq A.$$

A class which is a element of another class is called a *set*. We will let $x, y, z, \ldots$ be informal variables which range over sets. In many set theories (in particular ZFC) every class is already a set, but this is not always the case, e.g., the theories BGC^+ and BG^-C^+ discussed below.

In order to encode ordered pairs, triples, and other longer tuples in one's universe of sets an *axiom of pairing* is needed. This says that for any sets x and y there is a set z whose members are just x and y. Symbolically:

$$(\exists z)(\forall w)[w \in z \leftrightarrow (w = x \vee w = y)].$$

With the axiom of extensionality such z is unique and is informally written as $\{x, y\}$ (with $\{x\}$ also used to abbreviate $\{x, x\}$). The encoding of ordered pairs and other tuples now becomes

$$< x, y > = \{\{x\}, \{x, y\}\}$$
$$< x_1, x_2, \ldots, x_n > = << x_1, x_2, \ldots, x_{n-1} >, x_n > .$$

Regardless of whether or not the theory at hand requires all classes to be sets, it will be convenient (following [13]) to introduce additional *informal classes*. For each formula $\varphi(x, A_1, \ldots, A_n)$ of the language **L** all of whose quantified variables are set variables and whose free variables are among the list $x, A_1, \ldots, A_n$ we introduce the informal class $\{x : \varphi(x, A_1, \ldots, A_n)\}$ with the understanding that

$$y \in \{x : \varphi(x, A_1, \ldots, A_n)\} \text{ iff } \varphi(y, A_1, \ldots, A_n) ,$$
$$B = \{x : \varphi(x, A_1, \ldots, A_n)\} \text{ iff } (\forall x)[x \in B \leftrightarrow \varphi(x, A_1, \ldots, A_n)] , \text{ and}$$
$$\{x : \varphi(x, A_1, \ldots, A_n)\} \in B \text{ iff for some set } y, \text{ both } y \in B$$
$$\text{and } y = \{x : \varphi(x, A_1, \ldots, A_n)\} \text{ are true.}$$

Note that the informal class $\{x : \varphi(x, A_1, \ldots, A_n)\}$ depends on the parameters $A_1, \ldots, A_n$ (which may also be informal). It is to be also remembered that informal classes are *relative* to the particular language **L** in which the current set theory is expressed. Occasionally, an informal class will exist as a set and this is expressed as

$$(\exists y)[y = \{x : \varphi(x, A_1, \ldots, A_n)\}] .$$

If the language has formal class variables one can use

$$(\exists B)[B = \{x : \varphi(x, A_1, \ldots, A_n)\}]$$

to assert an informal class's existence as an actual class. As these examples indicate, we may pretend that informal classes are terms in some language extending **L** where they denote potential classes. We shall freely write formulas using them, but never as a quantified variable. With this understanding, any such formula φ can be canonically rewritten as an equivalent **L**-formula φ' where mention of informal classes is eliminated.

Further (informal) operations on classes that suggest themselves are:

$$\{A, B\} = \{x : x = A \lor x = B\}$$
$$A \cap B = \{x : x \in A \,\&\, x \in B\}$$
$$A \cup B = \{x : x \in A \lor x \in B\}$$
$$A - B = \{x : x \in A \,\&\, x \notin B\}$$
$$A \times B = \{x : (\exists y, z)[x = < y, z > \,\&\, y \in A \,\&\, z \in B]\}$$
$$\bigcup A = \{x : (\exists y)[x \in y \,\&\, y \in A]\}$$
$$P(A) = \{x : x \subseteq A\}^1$$
$$Dom(A) = \{x : (\exists y)[< x, y > \in A]\}$$
$$Ran(A) = \{x : (\exists y)[< y, x > \in A]\}$$
$$\in |_A = \{x : (\exists y, z)[x = < y, z > \,\&\, y, z \in A \,\&\, y \in z]\}.$$

---

[1]I use $P(A)$ here instead of $\mathcal{P}(A)$ to distinguish a class (informal or not) which plays a powerset-like role within a set theory from that which might be an "actual" powerset.

We use A^2, A^3 etc. to abbreviate $A \times A$, $A^2 \times A$, etc. A class R is a *relation* if

$$(\forall z)[z \in R \rightarrow (\exists x, y)[z = < x, y >]].$$

A class F is a *function* if it is a relation and

$$(\forall x, y, z)[< x, y >\in F \& < x, z >\in F \rightarrow y = z].$$

Special classes are $V = \{x : x = x\}$ and $\emptyset = \{x : x \neq x\}$. For function F and class A, $F[A]$ is $Ran(F \cap [A \times V])$. If $x \in Dom(F)$ then $F(x)$ is that unique y such that $< x, y >\in F$.

The theory ZFC can now be described. Its language $\mathbf{L}(ZFC)$ only mentions sets and has membership $\in$ as its sole nonlogical constant. Using informal class variables as described above, ZFC's axioms are expressible as:

ZFC : *extensionality*: sets x, y with the same elements are equal
  pairing: for any sets x and y, $\{x, y\}$ is a set
  separation: for every class A and set x, $x \cap A$ is a set
  union: for every set x, $\bigcup x$ is a set
  power set: for every set x, $P(x)$ is a set
  infinity: there is a set x such that $\emptyset \in x$ and
    whenever $y \in x$ then $y \cup \{y\} \in x$ as well
  replacement: if F is a function then for every set x,
    $F[x]$ is a set
  regularity: if $x \neq \emptyset$ then $y \in x$ exists for which $y \cap x = \emptyset$
  choice: if $y \neq \emptyset$ is the case for every $y \in x$, then a
    function f exists (as a set) such that $Dom(f) = x$ and
    $f(y) \in y$ for all $y \in x$.

Note that the axioms of separation and replacement involve quantification over informal classes, or (more literally) quantification over formulas of the language $\mathbf{L}(ZFC)$. These are examples of axiom schemata. A consequence of these axioms is the following variation of the axiom of replacement:

  collection: For any class R and set x, there exists a set y
    such that $x \cap Dom(R) \subseteq Dom(y \cap R)$.

This is often used (as an axiom *schema*) instead of the replacement axiom in set theories where the axioms of choice and regularity are not both assumed.

There is a stronger form of the axiom of choice which is possible, namely *global* choice. It will be convenient for us to express this in terms of a global well-ordering of the class of all sets. A binary class relation C is a *linear order* of V if it satisfies

  transitivity: $< x, y >, < y, z >\in C$ imply that $< x, z >\in C$
  irreflexivity: $< x, y >\in C$ implies $x \neq y$
  trichotomy: for all sets x, y either $< x, y >\in C$,
    $< y, x >\in C$ or $x = y$.

The linear order C has *small initial segments* if

$$\text{for all } x, \ \{y : \ <y, x> \in C\} \text{ is a set.}$$

It is, furthermore, a *well-ordering* if

$$\text{for all } x \neq \emptyset \text{ , there exists } y \text{ such that } <y, x> \in C$$
$$\text{and } x \cap \{z : <z, y> \in C\} = \emptyset.$$

We extend ZFC to a set theory ZFC^+ with global choice by adding a new binary predicate symbol $\mathcal{C}$ to the language $\mathbf{L}(ZFC)$ and including with the axioms of ZFC the new axiom

> *global choice*: the class $\{<x, y> : \mathcal{C}(x, y)\}$ is a linear order
> of all sets which has small initial segments and is
> a well-ordering.

Of course, the original axiom of choice is now redundant. Also to be noted is that the axiom schemata of separation and replacement (or collection, if used instead) are now taken in the stronger sense: the informal classes mentioned may involve formulas using the new predicate symbol $\mathcal{C}$.

The set theory ZFC without the axiom of choice is written ZF. I will write ZF^-C^+, ZF^-C and ZF^- to indicate ZFC^+, ZFC and ZF without the axiom of regularity[2]. Informally speaking, the axiom of regularity insures the "nonexistence" of infinitely descending $\in$-chains of sets $x_1 \ni x_2 \ni x_3 \ni \ldots \ni x_n \ni \ldots$. Such phenomenon has been traditionally considered pathological by set theorists and was routinely eliminated by the axiom of regularity in set universes being offered to the greater mathematical public. Opinion concerning this has recently shifted[3]. In any case, as mentioned in Chapter 3, any "non"standard treatment of set theory beyond the fragment chosen by Robinson (i.e., superstructures) has, with the slightest level of saturation, a violation of the axiom of regularity.

For this reason, we shall begin discussion by looking at the world from a ZF^- point of view. Omitted details can be found in [**13**]. In the discussion which follows the axioms of ZF^- are assumed to hold when suitably expressed in some (possible extended) set theoretic language $\mathbf{L}$ which has $\in$ as binary predicate and uses formal class variables. Thus the axiom schemata of separation and collection are to be taken in the strong sense.

A class A is *transitive* if $x \subseteq A$ whenever $x \in A$. A class A is *regular* if each of its elements $x \in A$ is a subset in a transitive set z such that

$$\text{for each } w \subseteq z \text{ there is a } y \in w \text{ such that } y \cap w = \emptyset.$$

We let $U = \{x : x \text{ regular}\}$ be the class of regular sets. As a subclass of V, U is regular, transitive and $(U, U|_\in)$ forms a natural model of ZF. Some authors have called U the class of "usual" sets.

---

[2]In each case the axiom of replacement (or for ZF^-, the axiom of collection) is retained.

[3]And rightly so: See [**1**] and [**3**]. See also [**4**] for pioneering work on alternative axioms to actively replace regularity.

An *ordinal* is any set x such that

$$x \text{ is transitive, regular and linearly ordered by } \in |_x.$$

We write $On = \{x : x \text{ is an ordinal}\}$ for the class of all ordinals. In fact, each ordinal $x \in On$ is well-ordered by $\in |_x$, as is On by $\in |_{On}$. We let $\alpha, \beta, \gamma, \ldots$ be informal variables ranging over ordinals. We write $\alpha < \beta$ to mean $\alpha \in \beta$. An ordinal is literally the set of its predecessors. The first ordinal 0 is $\emptyset$. The *successor* $\alpha + 1$ of an ordinal α is $\alpha \cup \{\alpha\}$. Ordinals not of this form other than 0 are *limit* ordinals. By the axiom of infinity, limit ordinals exist and by the well-ordering of ordinals there exists a first, which is written as ω. The elements of ω are the *finite* ordinals (also known as the *natural numbers*). *Cardinal numbers* are ordinals α which (within the ZF^--universe) cannot be put into a one to one correspondence with a predecessor $\beta < \alpha$. If κ is a cardinal, we write κ^+ for the first cardinal number following κ in the On-ordering. A subclass $On' \subseteq On$ is *unbounded* if for every $\alpha \in On$ there exists $\alpha' \in On'$ such that $\alpha < \alpha'$. An analogous notion holds for *unbounded* subsets $x \subseteq \alpha$ in a limit ordinal α. The class $On' \subseteq On$ is *closed* if for any limit ordinal $\alpha \in On$ one has $\alpha \in On'$ whenever $\alpha \cap On'$ is unbounded in α. *Closed* subsets $x \subseteq \alpha$ in a limit ordinal α are similarly defined. The closed unbounded subclasses in On (or in a limit α) are closed under finite intersections.

Transfinite induction on ordinals is valid, namely,

> For any class D, if for all ordinals α one has $\alpha \in D$ whenever $\alpha \subseteq D$, then $On \subseteq D$.

This makes available inductive definitions by transfinite recursion:

> For every function $H : V \to V$, there is a unique function $F : On \to V$ such that $F(\alpha) = H(F|_\alpha)$ for each $\alpha \in On$.

Here, $F|_\alpha = F \cap [\alpha \times V]$ is the restriction of F to α.

Applications of this principle give canonical addition and multiplication on ordinals which are both associative and together obey a left distributive law. On the natural numbers these operations coincide with the usual ones. Another application of transfinite recursion is to define inductively the class $U(x)$ of sets *regular over* a set x. This is to consist of all sets y such that every descending $\in$-chain starting with y, if continued long enough, either ends with $\emptyset$ or contains an element of x. Defined by transfinite induction $U(x)$ is $\bigcup_{\alpha \in On} U_\alpha(x)$ where

$$U_\alpha(x) = x \cup \bigcup_{\alpha' < \alpha} P(U_{\alpha'}(x)).$$

Clearly, $U(\emptyset) = U$. We write U_α for $U_\alpha(\emptyset)$.

In general, we call an On-indexed sequence of sets $\{W_\alpha\}_{\alpha \in On}$ (i.e., a function $F : On \to V$ where $F(\alpha) = W_\alpha$) *increasing* if $W_\alpha \subseteq W_\beta$ for all $\alpha < \beta$, and *continuous* if, for all limit $\alpha \in On$,

$$W_\alpha = \bigcup_{\alpha' < \alpha} W_{\alpha'}.$$

Clearly, the $U_\alpha(x)$ form an increasing continuous sequence.

Important for later discussion will be the *reflection principle*. This is a kind of local transfer principle as applied to single formulas. Let $\varphi = \varphi(\mathbf{x}, \mathbf{A})$ be a formula written strictly in the current language $\mathbf{L}$ (i.e., without informal classes) all of whose quantified variables are set variables and whose free variables are among the set variables $\mathbf{x} = x_1, \ldots, x_n$ and class variables $\mathbf{A} = A_1, \ldots, A_m$. For each class W (which may be informal), we write $^W\varphi = {}^W\varphi(\mathbf{x}, \mathbf{A})$ for the formula obtained by restricting all quantifier to W, i.e., each $(\forall y)$ and $(\exists z)$ gets rewritten as $(\forall y \in W)$ and $(\exists z \in W)$ — which for later purposes we shall abbreviate as $(\forall^W y)$ and $(\exists^W z)$. If $W \subseteq W'$ are classes, we shall say (relative to a given choice of actual classes $\mathbf{A}$) that W *reflects* φ *for* W' if

> for all $\mathbf{p} = p_1, \ldots, p_n \in W$ one
> has $^W\varphi(\mathbf{p}, \mathbf{A})$ iff $^{W'}\varphi(\mathbf{p}, \mathbf{A})$.

For the reflection principle, we consider an increasing continuous On-indexed sequence of sets $\{W_\alpha\}_{\alpha \in On}$ whose union $\bigcup_{\alpha \in On} W_\alpha$ is the class W. The principle states that for fixed $\mathbf{A}$

> there exists a closed unbounded class $On' \subseteq On$
> of ordinals α for which W_α reflects φ for W.

Although the proof of the principle only requires ZF^- axioms, the axiom schema of collection plays a crucial role.

A useful extension of this reflection principle is the following. Suppose $\mathbf{L}_0$ is the language of a type of model theoretic structure whose nonlogical constants are finite in number. Call an $\mathbf{L}_0$-structure $\mathbf{A} = (A, \ldots)$ *small* if A is a set. If $\varphi(\mathbf{x})$ is a formula in the language $\mathbf{L}_0$ whose free variables are among $\mathbf{x} = x_1, \ldots, x_n$ and if $\mathbf{A}' = (A', \ldots)$ is an $\mathbf{L}_0$-structure homomorphically[4] extending the $\mathbf{L}_0$-structure $\mathbf{A} = (A, \ldots)$, then we say that $\mathbf{A}$ *reflects* $\varphi(\mathbf{x})$ *for* $\mathbf{A}'$ if

> for all $\mathbf{p} = p_1, \ldots, p_n \in A$, one has
> $\mathbf{A} \models \varphi(\mathbf{p})$ iff $\mathbf{A}' \models \varphi(\mathbf{p})$.

Now let $\{\mathbf{A}_\alpha\}_{\alpha \in On}$ be an homomorphically increasing continuous On-indexed sequence of small $\mathbf{L}_0$-structures whose limit (possibly non-small) is $\mathbf{A} = (A, \ldots)$. Then the *extended reflection principle* states that:

> there exists a closed unbounded class $On' \subseteq On$
> of ordinals α for which $\mathbf{A}_\alpha$ reflects $\varphi(\mathbf{x})$ for $\mathbf{A}$.

---

[4]Homomorphisms were defined in Section 2.1.

The proof of the extended principle uses the fact that assertions of the form $\mathbf{A} \models \varphi(\mathbf{p})$ or $\mathbf{A}_\alpha \models \varphi(\mathbf{p})$ are equivalent to assertions $^{A'}\varphi'(\mathbf{p})$ or $^{A'_\alpha}\varphi'(\mathbf{p})$ in the current language $\mathbf{L}$, where A' is an appropriate class based on the structure $\mathbf{A}$, and the A'_α are appropriate sets, based on the structures $\mathbf{A}_\alpha$, which increase continuously and have A' as their limit. .

There is a set theory alternative to ZFC and its variants which has been useful in foundational studies, namely Bernays-Gödel Set theory BG. Its language $\mathbf{L}(BG)$ has class variables $A, B, C, \ldots$ and has $\in$ as its sole nonlogical constant. Sets are defined as usual and informal set variables $x, y, z, \ldots$ are introduced as before. The axioms of BG are those of ZF except that the axiom schemata of separation and replacement are now expressed (using $\mathbf{L}(BG)$'s class variables) as single axioms and the theory is augmented by the axiom of *comprehension*, namely,

> every informal class $\{x : \varphi(x, A_1, \ldots, A_n)\}$ exists as an actual class.

In symbols, this is

$$(\forall A_1, \ldots, A_n)(\exists B)[B = \{x : \varphi(x, A_1, \ldots, A_n)\}].$$

It is to be noted that an axiom of global choice C^+ for BG doesn't require a extension of language for its expression. We let BG^-, BGC^+ and BG^-C^+ denote the obvious variants of Bernays-Gödel set theory.

Of course, the axiom of comprehension is really an axiom schema. This initially leaves the impression that BG must have infinitely many axioms, which is truly the case for ZF. However, each instance of the comprehension axiom can be interpreted as saying that the universe of classes is closed under a certain operation, namely, the one that assigns the class $B = \{x : \varphi(x, A_1, \ldots, A_n)\}$ to the classes $A_1, \ldots, A_n$. It turns out that there is a finite list of such operations (known as *Gödel operations*) whose compositions generate all of the rest. Thus finitely many axioms expressing the fact that the universe of all classes is closed under these few operations suffice to imply the full axiom schema of comprehension.

As a set theory, BG proves to be a convenient starting point for constructing alternative universes of foundational interest. We mention the work of Vopěnka and Hájek [**24**] as an early example. By weakening the axiom of separation they allow for subclasses of sets $A \subseteq x$ which are not themselves sets. These they call *semisets*. Our set theory EST, to be presented in Chapter 12, will involve similar entities, but the role they play will differ from that in the Vopěnka-Hájek theory.

In EST, the existence of a universal class consisting of all sets will not be assumed. The focus will instead be on certain actual classes ("universes") A which essentially model BG^-C^+. The latter's finite axiomatizability allows us to express this aspect of universes by a single formula within EST's language

$(= \mathbf{L}(BG))$. Such universes A will extend to other universes ("enlargments") $A \subseteq A'$ having special properties. In particular, we will want the inclusion

$$(A, \in |_A) \subseteq (A', \in |_{A'})$$

to reflect arbitrary *restricted* formulas. These are formulas $\varphi(\mathbf{x})$ in EST's language whose variables are all set variables and whose quantifiers are restricted, i.e., of the form $(\forall z \in y)$ or $(\exists z \in y)$. To be able to define such enlargements within EST's language, it will be important to remove the implicit quantification over restricted formulas.

For this, the Gödel operations will once again prove useful. Call a formula $\theta(\mathbf{z}, y)$ in $\mathbf{L}(ZFC)$ a ZF^--*operator* if it is a ZF^--theorem that to every choice of sets $\mathbf{z}$ there exists exactly one set y such that $\theta(\mathbf{z}, y)$. In this sense, each Gödel operation is a ZF^--operator. Moreover, the defining formula $\theta(\mathbf{z}, y)$ in each case can be taken to be restricted. Say the extension

$$(A, \in |_A) \subseteq (A', \in |_{A'})$$

preserves such a ZF^--operator if

$$(A, \in |_A) \text{ reflects } \theta(\mathbf{z}, y) \text{ for } (A', \in |_{A'}).$$

Clearly, ZF^--operators are closed under composition. Indeed, compositions of preserved ZF^--operators are also preserved.

Now it is a ZF^--theorem (see [13]) that for every restricted formula $\varphi(\mathbf{x})$ the ZF^--operator

$$\theta(\mathbf{z}, y) \text{ iff } y = \{\mathbf{x} \in \mathbf{z} : \varphi(\mathbf{x})\}$$

is a composition of Gödel operations. It quickly follows that the extension

$$(A, \in |_A) \subseteq (A', \in |_{A'})$$

reflects arbitrary restricted formulas exactly when it preserves the Gödel operations. Since these are finite in number, this is a property expressible in EST's language.

CHAPTER 10

Current "Non"standard Set Theories

I now give a unified description of the "non"standard set theories IST, NST, NS_1, and NS_2. To make comparisons among theories easier, I shall freely re-order, rename, and in certain cases, alter the statements of the axioms as originally presented by their authors. In each case the changes made are inessential.

The common philosophy of these theories is that there is the universe S of all "standard" sets used by conventional mathematicians. In this universe, the axioms of ZFC are assumed to hold. The universe S is also to be elementarily embedded in a larger universe I of "internal" sets. As a class, I is to be transitive. For Nelson, in IST, the universe of all sets stops there. For the other authors there are further sets, and the totality of all sets (V in our notation) is referred to as the universe of "external" sets.

We can presume a common language $\mathbf{L}(NZFC)$ for the four theories ($NZFC$ = "Non"standard ZFC). $\mathbf{L}(NZFC)$ shall result from $\mathbf{L}(ZFC)$ by the addition of two new unary predicate symbols: st (asserting "standardness") and int (asserting "internalness"). I shall make free use of informal classes as done in Chapter 9. Fixing notation, we let $S = \{x : st(x)\}$, $I = \{x : int(x)\}$, $Fin = \{x : x \text{ finite}\}$, and for class A, $^{\circ}A$ denotes the class $A \cap S$. The expression "x finite" abbreviates the formula $\varphi(x)$ in $\mathbf{L}(ZFC)$ which asserts that every (external) set of subsets of x has a $\subseteq$-maximal element[1].

In each of the theories the class S models a ZFC universe. Specifically,

> For each formula φ from $\mathbf{L}(ZFC)$ which
> is an axiom of ZFC, $^{S}\varphi$ is the case.

Also assumed is

> $S \subseteq I$ and as a class, I is transitive.

A link between the external (all sets) and standard universe is that where

---

[1] In the presence of the axiom of choice this is equivalent to saying that no $(1-1)$ function $f : x \rightarrow x$ exists for which $f[x] \subsetneq x$.

possible each class can be "standardized", that is,

> For any class A, if there exists standard x
> such that $\,^\circ A \subseteq \,^\circ x$, then such x can be chosen
> so that $\,^\circ A = \,^\circ x$.

This is equivalent to saying that the standard universe S satisfies the axiom of separation with respect to the full language $\mathbf{L}(NZFC)$.

The internal universe I is to elementarily extend the standard universe S. This is expressed as the "set theoretic transfer principle", namely,

> for any formula $\varphi(\mathbf{x})$ from $\mathbf{L}(ZFC)$, one has
> $(\forall^S \mathbf{x})[\,^S\varphi(\mathbf{x}) \leftrightarrow \,^I\varphi(\mathbf{x})]$.

Here $\mathbf{x} = x_1, \dots, x_n$ includes all variables free in $\varphi(\mathbf{x})$.

A further feature common to all four theories is some level of saturation for I as a $\mathbf{L}(ZFC)$-structure. Let $\varphi(x, y, \mathbf{z})$ be a formula from $\mathbf{L}(ZFC)$ whose free variables are among x, y and $\mathbf{z} = z_1, \dots, z_n$. For fixed elements $z_1, \dots, z_n \in I$ and subclass $D \subseteq I$, the family of classes

$$W_x = \{y \in I : \,^I\varphi(x, y, \mathbf{z})\} \text{ for } x \in D$$

may or may not satisfy the finite intersection property. That it might can be expressed within the language $\mathbf{L}(NZFC)$ as

$$(\forall^{Fin} d \subseteq D)(\exists^I y)(\forall^I x \in d) \,^I\varphi(x, y, \mathbf{z}).$$

Saturation for I means that if D is "small" and the finite intersection property holds (in the foregoing sense), then the intersection of all the W_x's is nonempty, or in $\mathbf{L}(NZFC)$

$$(\exists^I y)(\forall^I x \in D) \,^I\varphi(x, y, \mathbf{z}).$$

Thus, in each of the four theories the axiom of saturation has the form

$$(\forall^I \mathbf{z})(\forall^{small} D \subseteq I)[(\forall^{Fin} d \subseteq D)(\exists^I y)(\forall^I x \in d) \,^I\varphi(x, y, \mathbf{z})$$
$$\to (\exists^I y)(\forall^I x \in D) \,^I\varphi(x, y, \mathbf{z})].$$

The sense of a "small" D varies for the separate theories, but in each case this can be expressed as saying there exists a "typically small" class B and an onto function $F : B \to D$. The requirements are as follows:

> for IST: $B = S$ and $F : S \to S$ is the identity map
> for NST: $B = S$ and $F : S \to D$ is an external set
> for NS_1 and NS_2 : $B = \,^\circ s$ for some standard set $s \in S$ and
> $\quad F : \,^\circ s \to D$ is an external set.

Kawai, in NST, refers to his "small" D as being S-sized. Hrbáček, in NS_1 and NS_2, refers to his "small" D as being *standard sized*.

The remaining axioms of the four theories concern properties of the external universe. These theories diverge on this issue as follows:

IST: $V = I$ (no truly external sets exist)

NST: V satisfies ZF^-C relative to $\mathbf{L}(NZFC)$;
    $S, I \in V$ are sets; and $V = U(I)$ (all external
    sets are regular over I)

NS_1: V satisfies ZF^-C relative to $\mathbf{L}(NZFC)$ minus
    the axioms of power set and choice (the axiom
    of collection is used instead of replacement);
    and every external set x can be "standardized"
    (has $^\circ x = {}^\circ s$ for some standard $s \in S$)

NS_2: Same as for NS_1 except the axiom of collection
    is removed and the axioms of power set and
    choice are reinstated.

The NST axiom $V = U(I)$ asserts a "quasi"regularity for external sets and can be conservatively added to both Hrbáček's NS_1 and NS_2.

CHAPTER 11

Proofs Of Conservativity

In this chapter I shall show that each of the four "non"standard set theories described above is "safe" for conventional mathematicians. Technically, this means showing for any sentence φ in $\mathbf{L}(ZFC)$ that $^S\varphi$ is a theorem in any one of these theories when and only when φ is already a theorem in ZFC. In this sense, the theories are *conservative* extensions of ZFC.

I first need to fix some notation and concepts. If T is one of the several theories with which we shall be dealing and $\mathbf{L}(T)$ is its language, then we write $T \vdash \varphi$ when φ is a sentence in $\mathbf{L}(T)$ which is provable in T. If $\mathbf{A}$ is a $\mathbf{L}(T)$-structure (more briefly, a T-structure), we write $\mathbf{A} \models \varphi$ when φ is true (is satisfied) in $\mathbf{A}$. If Φ is a collection of sentences in $\mathbf{L}(T)$, we write $T \vdash \Phi$ if $T \vdash \varphi$ for each $\varphi \in \Phi$. $\mathbf{A} \models \Phi$ shall have a similar meaning. Our most basic theory is Zermelo Set Theory or Z, and its structures are pairs (X, E) where X is a set and $E \subseteq X^2$ is a binary relation. For any $p \in X$, we shall write $|p|_E$ (called the E-*extension* of p) to mean $\{q \in X : qEp\}$. We say a subset $W \subseteq X$ is E-*represented* by the element $p \in X$ if it coincides with $|p|_E$. The Z-structure (X, E) is *extensional* if for any $p, q \in X$, $|p|_E = |q|_E$ implies $p = q$. Another Z-structure (X', E') *includes* (X, E) (written $(X, E) \subseteq (X', E')$) if $X \subseteq X'$ and $E \subseteq E'$, i.e., there is a homomorphic inclusion map of the structures. We say the inclusion of Z-structures is *transitive* if $|p|_E = |p|_{E'}$ for any $p \in X$. We say a set X is *transitively included* in a set X' if there is a transitive inclusion $(X, \in |_X) \subseteq (X', \in |_{X'})$ of Z-structures. If X is transitively included in the overall set theoretic universe V, we say that X itself is *transitive*. Note this agrees with previous terminology, namely, that X is transitive if for all $p \in X$, $p \subseteq X$. An extensional Z-structure (X, E) has a *set realization* if it has an isomorphic copy of the form $(X', \in |_{X'})$, for some transitive set X'.

The proofs of conservativity will be performed in a suitably flexible extension of ZF^-C^+ which I will call AST for **A**mbient **S**et **T**heory. The language $\mathbf{L}(AST)$ shall be gotten from $\mathbf{L}(Z)$ by adding a binary predicate symbol $\mathcal{C}$ to express the axiom of global choice (as before we let $C = \{< x, y >: \mathcal{C}(x, y)\}$) and an individual constant symbol V_0 to indicate a distinguished set in the AST-universe. The axioms of AST are those of ZF^-C^+ with respect to the full language $\mathbf{L}(AST)$ and additionally

AST :   *superuniversality* : If $(X, E) \subseteq (X', E')$ is a transitive
                inclusion of extensional Z-structures, then each
                isomorphism from (X, E) to the structure $(W, \in |_W)$
                on a transitive set W can be extended to an isomorphism
                from (X', E') to the structure $(W', \in |_{W'})$ on some
                transitive set $W \subseteq W'$

        elementary submodel : V_0 is a transitive set which is closed
                under formation of power sets and for which the
                structure $\mathbf{V}_0 = (V_0, \in |_{V_0}, C|_{V_0})$ is an elementary submodel
                of the entire $ZF^- C^+$ universe $\mathbf{V} = (V, \in, C)$.

Of course, the latter axiom is an axiom schema. I shall write On_0 for $On \cap V_0$
and U_0 for $U \cap V_0$ (note that $On_0 \in On$). Letting $\mathbf{V}_0(V_0)$ denote the structure
obtained from $\mathbf{V}_0$ by adding all elements $p \in V_0$ as distinguished individuals, I
call any $\mathbf{V}_0(V_0)$-basis set a V_0-*class*. If this fails to be an element of V_0, it is a
proper V_0-class; otherwise, it is a V_0-*set*. Of course, the typical V_0-class A is of
the form

$$A = \{x \in V_0 : \mathbf{V}_0 \models \varphi(x, \mathbf{p})\}$$

where $\varphi(x, \mathbf{z})$ is a formula of $\mathbf{L}(ZF^- C^+)$ and $\mathbf{p}$ is an appropriate $\mathbf{z}$-ary tuple of
elements from V_0.

  Some motivation for these new axioms is in order: In constructing models of
the "non"standard set theories IST, NST, NS_1, and NS_2 it is convenient to
have available a model of conventional set theory which is a *set*. Essentially, this
is the role played by V_0 in the elementary submodel axiom.

  When initially formed, our models of IST, etc. will usually be small struc-
tures $\widetilde{\mathbf{X}}' = (\widetilde{X}', \widetilde{E}', \dots)$ where $(\widetilde{X}', \widetilde{E}')$ is an underlying extensional Z-structure.
Rather than have to talk about $\widetilde{\mathbf{X}}'$-ordered pairs, $\widetilde{\mathbf{X}}'$-cartesian products and
the like, it will be convenient to pass to a set realization $(X', \in |_{X'})$ where X'
is a transitive set and be able to talk about ordinary ordered pairs, cartesian
products, and the like. Helpful to this program is the following fact provable in
ZF^-:

            Let $\varphi(\mathbf{x})$ be any restricted formula and let X be a
            transitive class. Then X reflects $\varphi(\mathbf{x})$ for the entire
            set universe V.

Since ordered pairs, cartesian products, and the like are definable by restricted
formulas, this means that the isomorphism of $(\widetilde{X}', \widetilde{E}')$ to $(X', \in |_{X'})$ sends
$\widetilde{\mathbf{X}}'$-ordered pairs, $\widetilde{\mathbf{X}}'$-cartesian products, and the like to X'-ordered pairs, X'-
cartesian products, and the like and that these are items in the ordinary sense.

  Of course, we need the *existence* of a set realization of $(\widetilde{X}', \widetilde{E}')$ for things to
work. If the Z-structure $(\widetilde{X}', \widetilde{E}')$ is "well-founded" in the sense that no infinite
descending $\widetilde{E}'$-chain exists, there is the folkloric Mostowski Collapsing Lemma

(see [13]) which asserts the existence of *exactly one* set realization $(X', \in |_{X'})$. In this case, X' is necessarily regular (i.e., an element of U). By the uniqueness asserted, the lemma easily extends to well-founded Z-structures $(\widetilde{X}', \widetilde{E}')$ where $\widetilde{X}'$ may be a proper class, provided for each $p' \in \widetilde{X}'$ that $|p'|_{\widetilde{E}'}$ is a set. We will refer to this extension as the *global* form of the Mostowski Collapsing Lemma.

The problem for us is that we already know that our $(\widetilde{X}', \widetilde{E}')$'s will *fail* well-foundedness since nonregularity in the external universe is forced. Thus, a new axiom needs to be added to ZF^- to insure the adequate existence of set realizations. It was Boffa (see [1] for a discussion) who first explored such extensions of ZF^- and the axiom of superuniversality, which we include amongst AST's axioms, is due to him. We will not need to exploit its full strength until Chapter 13. Until then we will rely on its weaker consequence, namely, that *every* small extensional Z-structure (X, E) has a set representation.

The following result is central to AST's usefulness.

THEOREM 11.1. *AST is a conservative extension of ZFC in the sense that if φ is any sentence of $\mathbf{L}(ZFC)$, then $ZFC \vdash \varphi$ iff $AST \vdash {}^U\varphi$.*

PROOF. The left right implication is clear. To see the reverse, suppose $AST \vdash {}^U\varphi$ and let AST_0 be a finitely axiomatized subtheory in which the proof takes place. According to [8], ZFC^+ is a conservative extension of ZFC, so we need only show that $ZFC^+ \vdash \varphi$. In ZFC^+ certain proper classes X, E and C can be inductively defined (consult [1]), so that one has

$$ZFC^+ \vdash \text{``}(X, E, C) \models AST_0{}^{*\text{''}}$$

where $AST_0{}^*$ consists of all axioms of AST_0 which don't refer to V_0 and where X, E, and C are interpreted as sets, membership and choice, respectively. By the extended reflection principle, we get

$$ZFC^+ \vdash (\exists v_0 \in X)[\text{``}(X, E, C, v_0) \models AST_0\text{''}]$$

(where v_0 is interpreted as V_0) and thus, that

$$ZFC^+ \vdash (\exists v_0)[\text{``}(X, E, C, v_0) \models {}^U\varphi\text{''}].$$

Since φ is in $\mathbf{L}(ZFC)$, this gives

$$ZFC^+ \vdash \text{``}(X, E) \models {}^U\varphi\text{''},$$

and if we let (in the ZFC^+ universe)

$$\widehat{U} = \{x \in X : x \text{ is } (X, E)\text{-well-founded}\}$$
$$\widehat{E} = E|_{\widehat{U}}$$

this further implies that

$$ZFC^+ \vdash \text{``}(\widehat{U}, \widehat{E}) \models \varphi\text{''}.$$

In the construction of X and E it can be assumed (see [1]) that both

$$ZFC^+ \vdash (\forall a \subseteq X)(\exists p \in X)[\text{``}p\ E\text{-represents } a\text{''}]$$
$$ZFC^+ \vdash (\forall a \in X)[\{q \in X : qEp\} \text{ is a set}]$$

and hence, by ZFC^+-collection, $\widehat{U}$'s largeness, by regularity and Mostowski's Collapsing Lemma (global form), we get that

$$ZFC^+ \vdash \text{``}(\widehat{U}, \widehat{E}) \text{ is isomorphic to } (V, \in)\text{''}$$

where V is the entire ZFC^+ universe. From this it follows easily that $ZFC^+ \vdash \varphi$. □

Our method for proving various "non"standard set theories are conservative over ZFC shall be automatic and uniform. It works as follows: Let the relevant "non"standard set theory be denoted by N and, as before, set $S = \{x : x \text{ is standard}\}$ and $I = \{x : x \text{ is internal}\}$. The axioms of N require that its standard part $(S, \in |_S)$ be a model of ZFC. Thus, for any sentence φ in $\mathbf{L}(ZFC)$ the implication

$$ZFC \vdash \varphi \Rightarrow N \vdash {}^S\varphi$$

is routine. Now in each case, the model of N we shall construct in the ambient universe will have $(S, \in |_S) \approx (U_0, \in |_{U_0})$. Thus, for any sentence φ in $\mathbf{L}(ZFC)$ we get, arguing backwards:

$$N \vdash {}^S\varphi \Rightarrow AST \vdash \text{``}(S, \in |_S) \models \varphi\text{''}$$
$$\Rightarrow AST \vdash {}^{U_0}\varphi \Rightarrow AST \vdash {}^U\varphi$$
$$\Rightarrow ZFC \vdash \varphi,$$

where the last implication uses Theorem 11.1.

I now proceed with construction of our models. Each shall be a $\mathbf{L}(NZFC)$-structure in the AST-universe of the form $\mathbf{X}' = (X', \in |_{X'}, X, {}^\circ X)$, where X' is the model's external universe and X and ${}^\circ X$ are, respectively, the model's internal and standard universes. We shall write $\mathbf{X}$ and ${}^\circ\mathbf{X}$ to denote $(X, \in |_X)$ and $({}^\circ X, \in |_{{}^\circ X})$, respectively. These are the model's *internal* and *standard parts*.

We begin by considering internal domains X for which the following are true:

as a set, X is transitive;
$\in |_X \subseteq X^2$ is clopen and
for some $\lambda \in On$, $({}^\circ X, \in |_{{}^\circ X}) \approx (U_\lambda, \in |_{U_\lambda})$.

It follows from the Mostowski Collapsing Lemma that the ordinal λ and the isomorphism of Z-structures are unique. For $\alpha \leq \lambda$, we let ${}^\circ X(\alpha)$ correspond to U_α under this isomorphism and define

$$X(\alpha) = \bigcup_{\alpha' < \alpha} \overline{{}^\circ X(\alpha' + 1)} \ .$$

If for each $\alpha < \lambda$, $X(\alpha + 1)$ is compact and ω-saturated (by construction it is already clopen), we will call X an $NZFC$-*domain* (of *order* λ). A routine argument in this case shows, for each $\alpha \leq \lambda$, that $X(\alpha)$ is also transitive and hence an $NZFC$-domain, now of order α.

There is a natural notion for *morphisms* between $NZFC$-domains. Let X and Y be $NZFC$-domains of orders λ and κ. A mapping $i : X \to Y$ will qualify as a morphism of the $NZFC$-domains if it is an ordinary morphism of the internal domains and also, if $\lambda \leq \kappa$ and the map i induces an isomorphism

$$({}^{\circ}X, \in |_{\circ X}) \approx ({}^{\circ}Y(\lambda), \in |_{\circ Y(\lambda)}).$$

It is routine to check (use Proposition 7.3 and superuniversality) that $NZFC$-domains are closed under direct limits.

To any $NZFC$-domain X of whatever order, we associate various $\mathbf{L}(NZFC)$-structures $\mathbf{X}' = (X', \in |_{X'}, X, {}^{\circ}X)$ which will be appropriate for modeling the theories IST, NST, NS_1 and NS_2. Three of these structures are simply described, namely,

$$\text{for modeling } IST \text{ set } X' = X$$
$$\text{for modeling } NST \text{ set } X' = U(X)$$
$$\text{for modeling } NS_2 \text{ set } X' = \text{the union}$$
$$\text{of all } U_\omega(p) \text{ for } p \in X.$$

The case for NS_1 is more subtle. We assume λ is the order of X and that $X = X(\lambda)$. For $\alpha \leq \lambda$, we inductively define $X'(\alpha)$ and the $NZFC$-structure $\mathbf{X}'(\alpha) = (X'(\alpha), \in |_{X'(\alpha)}, X(\alpha), {}^{\circ}X(\alpha))$ by specifying that

$$X'(0) = \emptyset; \text{ for limit } \alpha \leq \lambda, \ X'(\alpha) = \bigcup_{\alpha' < \alpha} X'(\alpha'); \text{ and}$$
$$\text{for } \alpha < \lambda, \ X'(\alpha + 1) \text{ is the union of } X'(\alpha), X(\alpha + 1)$$
$$\text{and the collection of all } \mathbf{X}'(\alpha)(X'(\alpha))\text{-basis sets,}$$

where $\mathbf{X}'(\alpha)(X'(\alpha))$ is the structure obtained from $\mathbf{X}'(\alpha)$ by including all elements of $X'(\alpha)$ as distinguished individuals. Note that the typical element in $X'(\alpha + 1)$ of the latter type is

$$\{r \in X'(\alpha) : \mathbf{X}'(\alpha) \models \varphi(r, \mathbf{p})\}$$

where $\varphi(y, \mathbf{x})$ is a formula of $\mathbf{L}(NZFC)$ with free variables among the y, $\mathbf{x} = y, x_1, \ldots, x_n$, and $\mathbf{p} = p_1, \ldots, p_n$ is an appropriate tuple of parameters taken from $X'(\alpha)$. The structure $\mathbf{X}' = (X', \in |_{X'}, X, {}^{\circ}X)$ appropriate for modeling NS_1 is obtained by setting $X' = X'(\lambda)$.

As may be already evident, I shall from now on be reversing the order of considering Hrbáček's theories NS_1 and NS_2 as the former demands extra technicalities. A further example of this is in the following notion of "definability". We will call an $NZFC$-domain X of order $\lambda \leq On_0$ *definable* if the isomorphism

$$({}^{\circ}X, \in |_{\circ X}) \approx (U_\lambda, \in |_{U_\lambda})$$

extends to an isomorphism

$$(X, \in |_X) \approx (\widetilde{X}, \widetilde{E})$$

where $\widetilde{X}$ and $\widetilde{E}$ are (possibly proper) V_0-classes. If $\mathbf{X}' = (X', \in |_{X'}, X, {}^{\circ}X)$ is the associated $\mathbf{L}(NZFC)$-structure appropriate for modeling NS_1, as described above, we also say that $\mathbf{X}'$ is *definable* if there is a further extension to an isomorphism of $\mathbf{X}'$ to a structure $\widetilde{\mathbf{X}}' = (\widetilde{X}', \widetilde{E}', \widetilde{X}, U_\lambda)$ where $\widetilde{X}'$ and $\widetilde{E}'$ are also (possibly proper) V_0-classes.

Such definability brings with it a notion of large and small with respect to which a principle of collection is valid. For example, suppose $\mathbf{X}'$ is definable. Then we will refer to any relation $R \subseteq X'^n$ ($\subseteq X'$ when λ is a limit, since X' is transitive) as also being *definable* if, in the bijection $X' \to \widetilde{X}'$, R corresponds to a V_0-class $\widetilde{R} \subseteq \widetilde{X}'^n$. If $\widetilde{R}$ is a V_0-set, we say that R is *small*, otherwise we shall call R *large*. The following collection principle is now obvious:

> If $R \subseteq X'$ is definable and $W \subseteq X'$ is small, then a small $Z \subseteq X'$ exists so that $W \cap Dom(R) \subseteq Dom(Z \cap R)$.

We shall refer to this principle as V_0-*collection transfered to* X'. By abuse of notation we shall also call an λ-indexed sequence $\{W_\alpha\}_{\alpha < \lambda}$ of small $W_\alpha \subseteq X'$ *definable* if the induced sequence of the same sets is definable when each index $\alpha < \lambda$ is replaced by its corresponding ${}^{\circ}X$-ordinal in ${}^{\circ}X$.

We will additionally call morphism $i : X \to Y$ between $NZFC$-domains *definable* if both X and Y are definable and the induced map $\widetilde{i} : \widetilde{X} \to \widetilde{Y}$ is a V_0-class. Similarly, a system

$$\{\tau_{ij} : X_i \to X_j\}_{(i,j) \in \Gamma}$$

of definable morphisms will be called *definable* if there is a bijection $\Gamma \to \widetilde{\Gamma}$ to a V_0-class $\widetilde{\Gamma}$ so that, letting

$$\{\widetilde{\tau}_{ij} : \widetilde{X}_i \to \widetilde{X}_j\}_{(i,j) \in \widetilde{\Gamma}}$$

be the induced system of V_0-class maps indexed by $\widetilde{\Gamma}$, one has

$$\{(i, j, p, q) \in V_0^4 : (i, j) \in \widetilde{\Gamma} \text{ and } (p, q) \in G(\widetilde{\tau}_{ij})\}$$

is itself a V_0-class. Clearly, the direct limit of a definable directed system of morphisms of $NZFC$-domains results in a $NZFC$-domain which is also definable.

Our central task in this chapter is the proof of the following theorem:

THEOREM 11.2. *Given any $NZFC$-domain X of order $\lambda = On_0$, the following are true:*

1) *If X is compact and ω-saturated, then the associated $\mathbf{L}(NZFC)$-structure $\mathbf{X}' = (X', \in |_{X'}, X, {}^{\circ}X)$ appropriate for IST is a model of IST.*

2) *If X is compact and $card(U_0)^+$-saturated, then the associated* $\mathbf{L}(NZFC)$-*structure* $\mathbf{X}' = (X', \in |_{X'}, X, {}^\circ X)$ *appropriate for NST is a model of NST.*

3) *If $X = X(On_0)$ and X is locally $card(U_0)$-saturated, then the associated* $\mathbf{L}(NZFC)$-*structure* $\mathbf{X}' = (X', \in |_{X'}, X, {}^\circ X)$ *appropriate for NS_2 is a model of NS_2.*

4) *If $X = X(On_0)$ and X is the direct limit of a definable directed system*
$$\{i_{\alpha\beta} : X[\alpha] \to X[\beta]\}_{\alpha<\beta\in On_0}$$
of small $NZFC$-domains such that for each cardinal $\kappa \in On_0$, $X[\alpha+1]$ is κ-saturated and compact for all sufficiently large $\alpha \in On_0$, then the associated $\mathbf{L}(NZFC)$-*structure* $\mathbf{X}' = (X', \in |_{X'}, X, {}^\circ X)$ *appropriate for NS_1 is a model of NS_1.*

Furthermore, in all cases the appropriate $NZFC$-domains exist and thus so do the associated models of "non"standard set theory.

PROOF. The construction of the appropriate $NZFC$-domains is done as follows: Let κ be any cardinal number. Since U_0 with the discrete topology is trivially a local internal domain, there exists, by Theorem 7.5, a κ-saturated local internal domain $\widetilde{Y}$ with core ${}^\circ\widetilde{Y} = U_0$. Let $\widetilde{E}$ be the closure in $\widetilde{Y}^2$ of $\in |_{U_0}$. By transfer, $(\widetilde{Y}, \widetilde{E})$ is an extensional Z-structure. Applying the axiom of superuniversality, we get an isomorphic copy of this in the form $(Y, \in |_Y)$ where Y is a transitive set. Transfering to Y the local internal domain structure on $\widetilde{Y}$, we get that $\in |_Y$ is clopen in Y^2. If $\kappa = \omega$ and $X = Y$, then X is appropriate for IST. If $\kappa = card(U_0)^+$ and $X = Y$, then X is appropriate for NST. If $\kappa = card(U_0)$ and $X = Y(On_0)$, then X is appropriate for NS_2.

For an X which is appropriate for NS_1, we use V_0-global choice to inductively define an increasing continuous On_0-indexed sequence $\{\widetilde{X}[\alpha]\}_{\alpha\in On_0}$ of local internal domains, each of which is a V_0-set, and such that for every $\alpha \in On_0$, $\widetilde{X}[\alpha]$ has core ${}^\circ\widetilde{X}[\alpha] = U_\alpha$ and $\widetilde{X}[\alpha+1]$ is compact, $card(\widetilde{X}[\alpha])^+$-saturated and contains $\widetilde{X}[\alpha]$ as an open set. Let $\widetilde{E}[\alpha]$ be the closure in $\widetilde{X}[\alpha]^2$ of $\in |_{U_\alpha}$ and let $(X[\alpha], \in |_{X[\alpha]})$ be a set realization of the extensional Z-structure $(\widetilde{X}[\alpha], \widetilde{E}[\alpha])$. Clearly, each $X[\alpha]$ is a small definable $NZFC$-domain and, for $\alpha < \beta \in On_0$, the inclusion $\widetilde{X}[\alpha] \subseteq \widetilde{X}[\beta]$ induces a definable morphism $i_{\alpha\beta} : X[\alpha] \to X[\beta]$ of $NZFC$-domains. Indeed, the entire system of morphisms of $NZFC$-domains

$$\{i_{\alpha\beta} : X[\alpha] \to X[\beta]\}_{\alpha<\beta\in On_0}$$

is definable and directed. Letting X be the $NZFC$-domain which is the direct limit of this system, we get an X which is appropriate for NS_1.

——PRELIMINARY AXIOMS[1]——

Assume now that X is any one of these appropriate $NZFC$-domains and that $\mathbf{X}' = (X', \in |_{X'}, X, {}^\circ X)$ is the associated $\mathbf{L}(NZFC)$-structure. We argue that

---

[1]Bold faced breaks such as this and those that follow are intended as sign posts for the reader to indicate the ongoing proof's overall structure.

$\mathbf{X}'$ models its appropriate "non"standard set theory. The following are routine observations:

$$°\mathbf{X}' = (°X, \in \, |_{°X}) \approx (U_0, \in \, |_{U_0})$$
$$\mathbf{X}' \models {}^S\varphi \text{ for each formula } \varphi \text{ of } \mathbf{L}(ZFC) \text{ which is an axiom of } ZFC$$
$$\mathbf{X}' \models S \subseteq I$$
$$\mathbf{X}' \models I \text{ is transitive.}$$

Since *any* subset of some $p \cap °X$ for $p \in °X$ is $(\in |_{°X})$-representable by an element $p' \in °X$, we have also that

all appropriate $\mathbf{X}'$-classes can be "standardized".

In the case of Hrbáček's NS_2 and NS_1, an inductive arguement shows that every external set $p \in X'$will lie in some $U_\alpha(X(\alpha))$ with $\alpha < \lambda$. A further inductive arguement shows that $p \cap X \subseteq X(\alpha + \alpha)$ and, since $X(\alpha + \alpha) \in °X$ when α is a nonlimit (which we may assume), it is clear that Hrbáček's "standardization" requirements for external sets are being met as well.

———SET THEORETIC TRANSFER PRINCIPLE———

Since X is compact for the IST and NST models (and hence is a *local* internal domain), the set theoretic transfer principle

for any formula $\varphi(\mathbf{x})$ from $\mathbf{L}(ZFC)$
$$\mathbf{X}' \models (\forall^S \mathbf{x})[{}^S\varphi(\mathbf{x}) \leftrightarrow {}^I\varphi(\mathbf{x})] \text{ is the case}$$

follows from the transfer principle for local internal domains. This argument is not available in the models for NS_1 and NS_2 since $X = X(On_0)$ is no longer a local set. Instead of this we will be able to use the ZFC axiom of collection and argue for every formula $\varphi(\mathbf{x}) \in \mathbf{L}(ZFC)$ with $\mathbf{x}$ n-ary that

$$(\star) \quad \{\mathbf{p} \in X^n : (X, \in |_X) \models \varphi(\mathbf{p})\} = \overline{\{\mathbf{p} \in °X^n : (°X, \in |_{°X}) \models \varphi(\mathbf{p})\}}.$$

The set theoretic transfer principle will then follow, since for any n-ary tuple $\mathbf{p}$ with entries in $°X^n$, we will have

$$\mathbf{X}' \models {}^S\varphi(\mathbf{p}) \Leftrightarrow (°X, \in |_{°X}) \models \varphi(\mathbf{p})$$
$$\Leftrightarrow \mathbf{p} \in °X^n \cap \overline{\{\mathbf{p}' \in °X^n : (°X, \in |_{°X}) \models \varphi(\mathbf{p}')\}}$$
$$\Leftrightarrow \mathbf{p} \in °X^n \cap \{\mathbf{p}' \in X^n : (X, \in |_X) \models \varphi(\mathbf{p}')\}$$
$$\Leftrightarrow (X, \in |_X) \models \varphi(\mathbf{p}) \Leftrightarrow \mathbf{X}' \models {}^I\varphi(\mathbf{p}).$$

I argue the case for $(\star)$ by induction on the complexity of $\varphi(\mathbf{x})$. For atomic formulas $x \in y$ and $x = y$ the issue is trivial. The induction case where φ is a boolean combination of simpler formulas is also routine. Since $\forall$ is expressible as $\neg \exists \neg$, we need only consider the case where $\varphi(\mathbf{x})$ is $(\exists y)\psi(y, \mathbf{x})$ and the condition $(\star)$ is known to hold for $\psi(y, \mathbf{x})$. Let

$$Z = \{(q, \mathbf{p}) \in °X^{n+1} : (°X, \in |_{°X}) \models \psi(q, \mathbf{p})\}.$$

By $(\star)$ for $\psi(y, \mathbf{x})$, we have

$$\overline{Z} = \{(q, \mathbf{p}) \in X^{n+1} : (X, \in |_X) \models \psi(q, \mathbf{p})\}.$$

Let $\pi : X^{n+1} \to X^n$ be the projection omitting the first coordinate. We are done if we can show $\pi[\overline{Z}] = \overline{\pi[Z]}$. Since the models for NS_1 and NS_2 have $X = X(On_0)$, X^n is covered by the compact clopens $X(\alpha)^n$ where $\alpha \in On_0$ is a *non*limit ordinal. Thus, it suffices to show

$$\pi[\overline{Z}] \cap X(\alpha)^n = \overline{\pi[Z]} \cap X(\alpha)^n$$

for all such α. The left to right inclusion is trivial since π is a continuous map. For the opposite inclusion we note $^\circ X(\alpha)^n$ is $\in |_{\circ X}$-represented[2] by some element of $^\circ X$. By the ZFC axiom of collection for $(^\circ X, \in |_{\circ X})$, there will exist $\alpha \leq \alpha' \in On_0$ (which we can take to be a nonlimit) such that

$$\pi[Z \cap {}^\circ X \times {}^\circ X(\alpha)^n] = \pi[Z \cap {}^\circ X(\alpha') \times {}^\circ X(\alpha)^n].$$

Since $Z \cap {}^\circ X(\alpha') \times {}^\circ X(\alpha)^n \subseteq {}^\circ X(\alpha')^{n+1}$ is local, it follows then that

$$\begin{aligned}
\overline{\pi[Z]} \cap X(\alpha)^n &= \overline{\pi[Z]} \cap \overline{{}^\circ X(\alpha)^n} = \overline{\pi[Z] \cap {}^\circ X(\alpha)^n} \\
&= \overline{\pi[Z \cap {}^\circ X \times {}^\circ X(\alpha)^n]} = \overline{\pi[Z \cap {}^\circ X(\alpha') \times {}^\circ X(\alpha)^n]} \\
&= \pi[\overline{Z \cap {}^\circ X(\alpha') \times {}^\circ X(\alpha)^n}] = \pi[\overline{Z} \cap X(\alpha') \times X(\alpha)^n] \\
&\subseteq \pi[\overline{Z} \cap X \times X(\alpha)^n] = \pi[\overline{Z}] \cap X(\alpha)^n,
\end{aligned}$$

and the desired equality holds.

Thus, for each model we have that $\mathbf{X}'$ satisfies the set theoretic transfer principle for $\mathbf{L}(NZFC)$-structures.

————SATURATION————

Next we take up the subject of saturation. Let $\varphi(x, y, \mathbf{z})$ be a formula from $\mathbf{L}(ZFC)$ whose free variables are among x, y and $\mathbf{z} = z_1, \ldots, z_n$. I assume we are given an appropriate "small" subclass $D \subseteq X$ and fixed elements $\mathbf{p} = p_1, \ldots, p_n \in X$ for which

$$\mathbf{X}' \models (\forall^{Fin} d \subseteq D)(\exists^I y)(\forall^I x \in d) \, {}^I\varphi(x, y, \mathbf{p})$$

is the case. Let $D' = D \cup \{p_1, \ldots, p_n\}$ and let $\mathbf{X}(D')$ be the structure obtained from $\mathbf{X} = (X, \in |_X)$ by adjoining the elements of D' as constants. Then the sets

$$W_r = \{q \in X : \mathbf{X} \models \varphi(r, q, \mathbf{p})\} \text{ for } r \in D$$

form a family of $\mathbf{X}(D')$-basis sets. To show set theoretic saturation, we need to show

$$\mathbf{X}' \models (\exists^I y)(\forall^I x \in D) \, {}^I\varphi(x, y, \mathbf{p}),$$

which amounts to showing that all the W_r's have a nonempty intersection.

I first argue that as a family of sets, the W_r's have the finite intersection property: Let $\mathbf{r} = r_1, \ldots, r_n \in D$ be a finite collection. Since $\mathbf{X} = (X, \in |_X)$ is

---

[2]i.e., for some $p \in {}^\circ X$, $^\circ X(\alpha)^n = |p|_{\in |_{\circ X}} = \{q :< q, p >\in (\in |_{\circ X})\}$

a model of ZFC there is an $\mathbf{X}$-set r' which $(\in|_X)$-represents $\{r_1, \ldots, r_n\}$, and since X is transitive and the formula in $\mathbf{L}(ZFC)$ expressing $x = \{r_1, \ldots, r_n\}$ is restricted, it follows that $r' = \{r_1, \ldots, r_n\}$. Clearly,

$$\mathbf{X}' \models \text{``r' is finite and $r' \subseteq D$''}$$

and our assumptions imply that the W_r's corresponding to $r_1, \ldots, r_n$ have a nonempty intersection.

We consider, case by case, the desired the nonemptyness of the intersection of all the W_r's. For IST we have $D = S = {}^\circ X$, so that $D' - {}^\circ X \subseteq \{p_1, \ldots, p_n\}$ is finite. Since by construction, X is compact and ω-saturated, and since $\in|_X$ is clopen we get the nonempty intersection desired. For the NST, NS_2 and NS_1 cases, we have $D \in X'$ and for some elements B, $F \in X'$, we further have

$$\mathbf{X}' \models \text{``$F : B \to D$ is an onto map''},$$

where $B \subseteq {}^\circ X$ is "typically" small. Since X' is transitive and $F : B \to D$ can be expressed in $\mathbf{L}(ZFC)$ by a restricted formula, we have that F is indeed an onto map from B to D. This means that $\operatorname{card}(D) \leq \operatorname{card}(B)$ and hence, $\operatorname{card}(D') \leq \operatorname{card}(B)$, since B can be assumed to be infinite. In the case of NST, $B = S = {}^\circ X$ and since by construction X is $\operatorname{card}({}^\circ X)^+$-saturated as local internal domain, we again get a nonempty intersection of the W_r's.

In the cases of NS_2 and NS_1, we have $B = {}^\circ p$ for some $p \in {}^\circ X$ so that $\operatorname{card}(D') < \operatorname{card}({}^\circ X)$. However, it is not clear that the W_r's are local sets and, in the case of NS_1, it is not even certain that X is $\operatorname{card}(D')^+$-saturated. What we really need for both cases is some compact $\operatorname{card}(D)^+$-saturated subinternal domain $\widetilde{X} \subseteq X$ which contains D' and for which the collection of

$$W_r \cap \widetilde{X} \text{ for } r \in D$$

also satisfies the finite intersection property. Letting $\widetilde{\mathbf{X}} = (\widetilde{X}, \in|_{\widetilde{X}})$, the property $(\star)$ will then insure that these sets are $\widetilde{\mathbf{X}}(D')$-basis sets which, by saturation, will have a nonempty intersection.

I now demonstrate the existence of such $\widetilde{X} \subseteq X$. Since $F \subseteq X^2 \subseteq X$ and $F \in X'$ it follows from previous remarks concerning Hrbáček standardization that $F = F \cap X$ must be contained in some $X(\alpha)$ for $\alpha \in On_0$. Since $X(\alpha)$ is transitive, we also have $D \subseteq X(\alpha)$. Using a possibly larger α, we can assume that $D' \subseteq X(\alpha)$. For any finite $d \subseteq X(\alpha)$ we clearly have

$$\mathbf{X} \models \text{``d is finite and $d \in X(\alpha + 1)$''}.$$

Since the set theoretic transfer principle holds, $\mathbf{X}$ inherits from ${}^\circ \mathbf{X}$ the ZF^-C axiom of collection, and this can be applied to show that for some (and hence all) sufficiently large $\beta \in On_0$, the collection

$$W_r \cap X(\beta) \text{ for } r \in D$$

also satisfies the finite intersection property. In the case of NS_2, we can take $\widetilde{X} = X(\beta + 1)$, since by construction this is $\text{card}(^\circ X)$-saturated. For the case of NS_1, we shall subsequently show not only that $\mathbf{X}$ is definable, but that *so is* $\mathbf{X}'$. Assuming this, we see that since $^\circ p$ is small and $F : {}^\circ p \to D$ is definable that V_0-collection transfered to X' implies that D is also small. Of course then so is D'. Letting $i_\alpha : X[\alpha] \to X$ for $\alpha \in On_0$ be the canonical morphisms into the inductive limit and writing $i_\alpha(X[\alpha])$ as $X\{\alpha\}$, we get, by two applications of transfered V_0-collection, first that for all large α, D' is contained in $X\{\alpha\}$ (which is $\text{card}(D')^+$-saturated when α is a nonlimit) and secondly, that again for large α the collection

$$W_r \cap X\{\alpha\} \text{ for r } \in D$$

also satisfies the finite intersection property. Thus, in the case of NS_1, we can put $\widetilde{X} = X\{\alpha\}$ for a sufficiently large nonlimit $\alpha \in On_0$. Therefore (modulo one gap in the case of NS_1), the proposed models for IST, NST, NS_2, and NS_1 all satisfy their corresponding saturation axioms.

——IST, NST, NS₂ CONCLUDED——

Except for the case of NS_1, the completion of model verifications is now routine. For IST there is nothing to check and for NST it is easily argued that $X' = U(X)$ models ZF^-C. The argument for this turns on the fact that the AST-universe models ZF^-C and that each AST-subset $p \subseteq U(X)$ is an element $p \in U(X)$. It is also easily checked for the NS_2 model that since X' is is transitive and closed under power sets, that it satisfies ZF^-C without the axiom of collection.

——MACHINERY TO FINISH NS₁——

We are left to verify three things concerning the proposed model for NS_1, namely, that

> $\mathbf{X}'$ is definable
> $\mathbf{X}'$ satisfies separation
> $\mathbf{X}'$ satisfies collection.

These issues will require us to return for a deeper study of $NZFC$-domains. From now on all such domains X will be assumed to satisfy $X = X(\lambda)$, where λ (now arbitrary) is the order of X. By $\mathbf{X}' = (X', \in |_{X'}, X, {}^\circ X)$, I shall automatically mean the associated $\mathbf{L}(NZFC)$-structure $\mathbf{X}'(\lambda)$ which is appropriate for modeling Hrbáček's NS_1. The nontrivial issues for us to investigate are whether definability of an $NZFC$-domain X guarantees the definability of the associated $\mathbf{L}(NZFC)$-structure $\mathbf{X}'$ and if so, does that also mean that a definable directed sytem of $NZFC$-domains will lift to a definable directed system of the associated $\mathbf{L}(NZFC)$-structures.

To answer these questions, we shall undertake a delicate study of model theoretic types of tuples $\mathbf{p} \in X'^n$. For our purposes, $\mathbf{p}$'s type as a tuple in $\mathbf{X}'$

will be too crude a classification. I shall refine these types by augmenting the structure $\mathbf{X}'$ to form a structure $\mathcal{X}'$ and then consider $\mathbf{p}$'s type as a tuple in $\mathcal{X}'$. We arrive at $\mathcal{X}'$ by adding to $\mathbf{X}'$ additional sets and relations. Assume X is of order λ. Then for each $\alpha < \lambda$, $X(\alpha)$ and $X'(\alpha)$ are subsets of X' and we include each as new unary relations in $\mathcal{X}'$. For each $n \geq 1$ and tuple $\mathbf{r} \in X^n$, we also include $\overline{\{\mathbf{r}\}} \subseteq X^n$ as a new n-ary relation in $\mathcal{X}'$. This completes the augmented structure. The language $\mathbf{L}_\lambda(NZFC)$ appropriate to $\mathcal{X}'$ can be described as follows: it consists of $\mathbf{L}(NZFC)$ along with unary predicate symbols χ_α and χ'_α, denoting $X(\alpha)$ and $X'(\alpha)$ for any $\alpha < \lambda$, and including also, for each $n \geq 1$, various n-ary predicate symbols $\mathcal{R}$, one for each ultrafilter on some $U^n_{\alpha+1}$ where both $\alpha < \lambda$ and $U^n_\alpha \notin \mathcal{R}$ hold. In the structure $\mathcal{X}'$, such predicate symbol $\mathcal{R}$ will denote $\overline{\{\mathbf{r}\}} \subseteq X(\alpha+1)^n$ $(= {}^\circ\overline{X(\alpha+1)^n})$ where $\mathbf{r} \notin X(\alpha)^n$ and the ultrafilter

$$\{R \subseteq {}^\circ X(\alpha+1)^n : \mathbf{r} \in \overline{R}\}$$

on ${}^\circ X(\alpha+1)^n$ corresponds, via the bijection ${}^\circ X(\alpha+1) \to U_{\alpha+1}$, to the ultrafilter $\mathcal{R}$ on $U^n_{\alpha+1}$. If Y is another $NZFC$-domain of order λ and this latter symbol $\mathcal{R}$ also denotes in $\mathcal{Y}'$ (augmentation of $\mathbf{Y}'$) some $\overline{\{\mathbf{s}\}} \subseteq Y(\alpha+1)^n$, I shall say that the tuples $\mathbf{r} \in X^n$ and $\mathbf{s} \in Y^n$ *simply correspond*.

An important observation concerning this notion of simple correspondence is that it is "ω-homogeneous". In other words, if $\mathbf{r} \in X^n$ and $\mathbf{s} \in Y^n$ are simply corresponding tuples then to each $m \geq 1$ and $\mathbf{r}_0 \in X^m$, there exists an $\mathbf{s}_0 \in Y^m$ such that the concatenated tuples $\mathbf{r}_0\mathbf{r} \in X^{m+n}$ and $\mathbf{s}_0\mathbf{s} \in Y^{m+n}$ are still simply corresponding. This is seen as follows: By local compactness a concatenated tuple $\mathbf{s}'_0\mathbf{s}' \in Y^{m+n}$ can be found which simply corresponds to $\mathbf{r}_0\mathbf{r} \in X^{m+n}$. But then $\mathbf{r} \in X^n$ corresponds simply to both $\mathbf{s}$, $\mathbf{s}' \in Y^n$ which implies that $\overline{\{\mathbf{s}\}} = \overline{\{\mathbf{s}'\}}$. Since Y is locally ω-saturated, it is also locally ω-homogeneous and we can find $\mathbf{s}_0 \in Y^m$ such that $\overline{\{\mathbf{s}_0\mathbf{s}\}} = \overline{\{\mathbf{s}'_0\mathbf{s}'\}}$. By definition, $\mathbf{r}_0\mathbf{r} \in X^{m+n}$ will correspond simply to this $\mathbf{s}_0\mathbf{s} \in Y^{m+n}$.

Another easy observation is the fact that if $r_1r_2 \in X^2$ corresponds simply to $s_1s_2 \in Y^2$, then $r_1 = r_2$ iff $s_1 = s_2$, and $r_1 \in r_2$ iff $s_1 \in s_2$. Indeed, let $\mathcal{R}$ name both $\overline{\{r_1r_2\}} \in X(\alpha+1)^2$ and $\overline{\{s_1s_2\}} \in Y(\alpha+1)^2$, for appropriate $\alpha < \lambda$. Then

$$r_1 = r_2 \Leftrightarrow \overline{\{r_1r_2\}} \subseteq \Delta_{X(\alpha+1)}\,(= \overline{\Delta_{{}^\circ X(\alpha+1)}}) \Leftrightarrow \Delta_{U(\alpha+1)} \in \mathcal{R}$$

and similarly,

$$s_1 = s_2 \Leftrightarrow \Delta_{U(\alpha+1)} \in \mathcal{R}$$
$$r_1 \in r_2 \Leftrightarrow \in|_{U(\alpha+1)} \in \mathcal{R}$$
$$s_1 \in s_2 \Leftrightarrow \in|_{U(\alpha+1)} \in \mathcal{R}.$$

We shall call a formula $\varphi(\mathbf{x})$ from $\mathbf{L}_\lambda(NZFC)$ *regular* if it does not contain any of the $\mathcal{R}$-symbols. On any $NZFC$-domain X of order λ, the $X'(\alpha)$ and $X(\alpha)$, for $\alpha < \lambda$, are all elements of X'. Thus, a regular $\varphi(\mathbf{x})$ is always $\mathcal{X}'$-equivalent to some $\varphi'(\mathbf{x}, \mathbf{p})$ where $\varphi'(\mathbf{x}, \mathbf{z})$ is a formula from $\mathbf{L}(NZFC)$ and $\mathbf{p}$ is a tuple of parameters from X'.

I now describe the key concept to be used in our analysis of $\mathcal{X}'$-types. Assume X and Y are appropriate $NZFC$-domains of the same order λ. We shall say that tuples $\mathbf{p} \in X'^n$ and $\mathbf{q} \in Y'^n$ *correspond* if there exist some simply corresponding $\mathbf{r} \in X^m$ and $\mathbf{s} \in Y^m$ such that for any pair p, q of matching entries from $\mathbf{p}$ and $\mathbf{q}$ it is the case that

> EITHER: p, q coincide with matching entries from $\mathbf{r}$ and $\mathbf{s}$,
> OR: for some $\alpha < \lambda$, there exists regular formula $\varphi(y, \mathbf{x})$ from
> $\quad \mathbf{L}_\alpha(NZFC)$ with $\mathbf{x}$ m-ary such that for each variable
> $\quad x$ in $\mathbf{x}$ that actually occurs free in $\varphi(y, \mathbf{x})$, the matching
> $\quad$ entries r, s from $\mathbf{r}$ and $\mathbf{s}$ lie in $X(\alpha)$ and $Y(\alpha)$, respectively,
> $\quad$ and both
> $$p = \{y \in X'(\alpha) : \mathcal{X}'(\alpha) \models \varphi(y, \mathbf{r})\}$$
> $$q = \{y \in Y'(\alpha) : \mathcal{Y}'(\alpha) \models \varphi(y, \mathbf{s})\}$$
> hold.

We shall speak of the simply corresponding tuples $\mathbf{r}$ and $\mathbf{s}$ as *generating* the correspondence between $\mathbf{p}$ and $\mathbf{q}$.

——KEY LEMMAS CONCERNING THE NS₁ MACHINERY——

First, some general observations concerning this notion of corresponding tuples.

LEMMA 11.1. *Let X and Y be $NZFC$-domains both of order λ. Then:*

a) *For $\alpha < \lambda$, any tuples $\mathbf{p}$ and $\mathbf{q}$ which correspond in $X'(\alpha)$ and $Y'(\alpha)$ continue to correspond in X' and Y'.*

b) *For any $p \in X'$, if $p \notin X$, then there exists some $\alpha < \lambda$, $\mathbf{r} \in X(\alpha)^m$ and regular formula $\varphi(y, \mathbf{x})$ from $\mathbf{L}_\alpha(NZFC)$ with $\mathbf{x}$ m-ary such that*
$$p = \{y \in X'(\alpha) : \mathcal{X}'(\alpha) \models \varphi(y, \mathbf{r})\}.$$

c) *For every tuple $\mathbf{p}$ from X', there exists a tuple $\mathbf{r}$ from X and regular formula $\psi(\mathbf{x}, \mathbf{z})$ from $L_\lambda(NZFC)$, where $\mathbf{x}$ and $\mathbf{z}$ have arities matching those of $\mathbf{p}$ and $\mathbf{r}$, such that*
$$\mathcal{X}' \models (\forall \mathbf{x})[\mathbf{x} = \mathbf{p} \leftrightarrow \psi(\mathbf{x}, \mathbf{r})].$$

d) *For every tuple $\mathbf{p}$ from X', there exists a tuple $\mathbf{q}$ from Y' which corresponds to $\mathbf{p}$.*

e) *If tuples $\mathbf{p}$ and $\mathbf{q}$ from X' and Y' have a correspondence generated by simply corresponding tuples $\mathbf{r}$ and $\mathbf{s}$ from X and Y and if $\mathbf{p}_0$ is a further tuple from X', then tuples $\mathbf{r}_0$ and $\mathbf{s}_0$ from X and Y can be found for which $\mathbf{r}_0\mathbf{r}$ and $\mathbf{s}_0\mathbf{s}$ also simply correspond and now generate a correspondence of $\mathbf{p}_0\mathbf{p}$ with some $\mathbf{q}_0\mathbf{q}$; furthermore, these choices can be arranged so that for each pair of matching entries p_0, q_0 from $\mathbf{p}_0$ and $\mathbf{q}_0$ either of the following two conditions can be imposed: 1) if $p_0 \in X(\alpha)$ for some $\alpha \leq \lambda$, then $q_0 \in Y(\alpha)$ and the entries p_0, q_0 coincide with* *matching entries from* $\mathbf{r}_0$ *and* $\mathbf{s}_0$ *or* 2) *if* $p_0 \in X'(\alpha)$ *for some* $\alpha < \lambda$, *then* $q_0 \in Y'(\alpha)$ *as well.*

PROOF. (a) is obvious. We prove (b) by induction on λ. We can assume $\lambda = \lambda_0 + 1$, that (b) is true for λ_0, and that $p \notin X'(\lambda_0)$. Thus, for some $\mathbf{p}' = p'_1, p'_2, \ldots, p'_m \in X'(\lambda_0)^m$ and formula $\psi(y, \mathbf{z})$ from $\mathbf{L}(NZFC)$ with $\mathbf{z}$ m-ary, we can write

$$p = \{y \in X'(\lambda_0) : \mathbf{X}'(\lambda_0) \models \psi(y, \mathbf{p}')\}.$$

Since (b) is true for λ_0, it easily follows that for each $i = 1, 2, \ldots, m$ there exist regular formulas $\psi_i(y, \mathbf{x}_i)$ from $\mathbf{L}_{\lambda_0}(NZFC)$ and appropriate tuple $\mathbf{r}_i$ from $X(\lambda_0)$ such that

$$p'_i = \{y \in X'(\lambda_0) : \mathcal{X}'(\lambda_0) \models \psi_i(y, \mathbf{r}_i)\}.$$

We can assume that there is no overlap among the variables occuring in the different $\mathbf{x}_i$. Form the concatenated tuples

$$\mathbf{x} = \mathbf{x}_1 \mathbf{x}_2 \ldots \mathbf{x}_m \text{ and } \mathbf{r} = \mathbf{r}_1 \mathbf{r}_2 \ldots \mathbf{r}_m$$

and let $\theta(\mathbf{z}, \mathbf{x})$ be the conjuntion, for $i = 1, 2, \ldots, m$, of the

$$(\forall w)[w \in z_i \leftrightarrow \psi_i(w, \mathbf{x}_i)],$$

where z_i is the ith entry of $\mathbf{z}$. Finally, let $\varphi(y, \mathbf{x})$ be

$$(\exists \mathbf{z})[\theta(\mathbf{z}, \mathbf{x}) \,\&\, \psi(y, \mathbf{z})].$$

Then $\varphi(y, \mathbf{x})$ is a regular formula from $\mathbf{L}_\lambda(NZFC)$ and

$$p = \{y \in X'(\lambda_0) : \mathcal{X}'(\lambda_0) \models \varphi(y, \mathbf{r})\}$$

holds. This proves (b).

We argue (c). Let $\mathbf{p} = p_1, p_2, \ldots, p_n \in X'^n$ and for each $i = 1, 2, \ldots, n$ choose tuple $\mathbf{r}_i$ from X and formula $\psi_i(x_i, \mathbf{z}_i)$ from $\mathbf{L}_\lambda(NZFC)$, with $\mathbf{z}_i$ having the arity of $\mathbf{r}_i$ and the x_i's being distinct, so that

if $p_i \in X$, then $\mathbf{r}_i = p_i$ (unary), $\mathbf{z}_i = z_i$ and $\psi_i(x_i, \mathbf{z}_i)$ is $x_i = z_i$;
if $p_i \notin X$, then $\psi_i(x_i, \mathbf{z}_i)$ is
$(\forall y)[y \in x_i \leftrightarrow \chi'_{\alpha_i}(y) \,\&\, " \mathcal{X}'(\alpha_i) \models \varphi_i(y, \mathbf{r}_i)"]$
where $\alpha_i < \lambda$ and $\varphi_i(y, \mathbf{r}_i)$ are derived from (b).

Let $\mathbf{z} = \mathbf{z}_1 \mathbf{z}_2 \ldots \mathbf{z}_n$ (we can assume the $\mathbf{z}_i$'s don't overlap) and let $\psi(\mathbf{x}, \mathbf{z})$ be the conjunction of the $\psi_i(x_i, \mathbf{z}_i)$ for $i = 1, 2, \ldots, n$. Then clearly, $\psi(\mathbf{x}, \mathbf{z})$ is regular and

$$\mathcal{X}' \models (\forall \mathbf{x})[\mathbf{x} = \mathbf{p} \leftrightarrow \psi(\mathbf{x}, \mathbf{r})],$$

where $\mathbf{r} = \mathbf{r}_1 \mathbf{r}_2 \ldots \mathbf{r}_n$.

In proving (d) we are free to permute the entries of $\mathbf{p}$ in any manner, so we can assume that $\mathbf{p}$ is a concatenation $\mathbf{r}' \mathbf{p}'$ where $\mathbf{r}'$ is a tuple from X and $\mathbf{p}' = p'_1, p'_2, \ldots, p'_m$ is a tuple from X', none of whose entries lie in X. By induction we can assume (d) is true for all $\alpha < \lambda$ and by (a) this means we can take $\lambda = \lambda_0 + 1$. By (b) we can find for each $i = 1, 2, \ldots, m$ a regular formula $\psi_i(y, \mathbf{x}_i)$ from $\mathbf{L}_{\lambda_0}(NZFC)$ and appropriate tuple $\mathbf{r}_i$ from $X(\lambda_0)$ such that

$$p_i' = \{y \in X'(\lambda_0) : \mathcal{X}'(\lambda_0) \models \psi_i(y, \mathbf{r}_i)\}.$$

Forming the concatenated tuples

$$\mathbf{x} = \mathbf{x}' \mathbf{x}_1 \mathbf{x}_2 \ldots \mathbf{x}_m \text{ and } \mathbf{r} = \mathbf{r}' \mathbf{r}_1 \mathbf{r}_2 \ldots \mathbf{r}_m$$

where $\mathbf{x}'$ is new and has arity matching that of $\mathbf{r}'$, we can also consider each $\psi_i(y, \mathbf{x}_i)$ as $\psi_i(y, \mathbf{x})$ so that

$$p_i' = \{y \in X'(\lambda_0) : \mathcal{X}'(\lambda_0) \models \psi_i(y, \mathbf{r})\}.$$

By local compactness, we can choose tuple $\mathbf{s} = \mathbf{s}' \mathbf{s}_1 \mathbf{s}_2 \ldots \mathbf{s}_m$ from Y which corresponds simply to $\mathbf{r}$. Then clearly, the simply corresponding tuples $\mathbf{r}$ and $\mathbf{s}$ generate the correspondence of $\mathbf{p}$ and $\mathbf{q} = \mathbf{s}' \mathbf{q}'$, where $\mathbf{q}' = q_1', q_2', \ldots, q_m'$ and for $i = 1, 2, \ldots, m$

$$q_i' = \{y \in X'(\lambda_0) : \mathcal{Y}'(\lambda_0) \models \psi_i(y, \mathbf{s})\}.$$

Finally, (e). Assume tuples $\mathbf{p}$ and $\mathbf{q}$ have a correspondence generated by the simply corresponding tuples $\mathbf{r}$ and $\mathbf{s}$. By induction, we can assume in proving (e) that the tuple $\mathbf{p}_0$ has only one entry, namely p_0. If p_0 lies in some $X(\alpha)$, then by the "ω-homogeneity" of simple correspondence we can choose q_0 in $Y(\alpha)$ so that $p_0 \mathbf{r}$ and $q_0 \mathbf{s}$ also simply correspond and these will clearly generate the correspondence of $p_0 \mathbf{p}$ and $q_0 \mathbf{q}$. If p_0 lies in some $X'(\alpha)$ but not in $X(\alpha)$, then by (b) find $\alpha' < \alpha$, $\mathbf{r}_0 \in X(\alpha')^m$ and regular formula $\varphi(y, \mathbf{x})$ from $\mathbf{L}_{\alpha'}(NZFC)$ with $\mathbf{x}$ m-ary such that

$$p_0 = \{y \in X'(\alpha') : \mathcal{X}'(\alpha') \models \varphi(y, \mathbf{r}_0)\}.$$

Use "ω-homogeneity" to find $\mathbf{s}_0 \in Y(\alpha')^m$ so that $\mathbf{r}_0 \mathbf{r}$ and $\mathbf{s}_0 \mathbf{s}$ also correspond simply. Then $\mathbf{r}_0 \mathbf{r}$ and $\mathbf{s}_0 \mathbf{s}$ clearly generate the correspondence of $p_0 \mathbf{p}$ and $q_0 \mathbf{q}$ where

$$q_0 = \{y \in Y'(\alpha') : \mathcal{Y}'(\alpha') \models \varphi(y, \mathbf{s}_0)\}.$$

In this latter case, q_0 clearly also lies in $Y'(\alpha)$. $\square$

The next lemma states the central fact about $\mathcal{X}'$-types. It is crucial information.

LEMMA 11.2. *Let X and Y be $NZFC$-domains both of order λ. Then tuple $\mathbf{p} \in X'^m$ and $\mathbf{q} \in Y'^m$ correspond if and only if they have the same $\mathbf{L}_\lambda(NZFC)$-type.*

PROOF. Suppose tuple $\mathbf{p}$ in $\mathcal{X}'$ and $\mathbf{q}$ in $\mathcal{Y}'$ have the same $\mathbf{L}_\lambda(NZFC)$-type. I shall argue that the tuples must be corresponding. Permuting entries does not alter matters, so that we can assume $\mathbf{p}$ is a concatenation $\mathbf{r}'\mathbf{p}'$, where $\mathbf{r}'$ is a tuple from X and $\mathbf{p}' = p_1', p_2', \dots, p_m'$ is a tuple from X', none of whose entries lies in X. Since types match, this means $\mathbf{q}$ is a similar concatenation $\mathbf{s}'\mathbf{q}'$ having the same properties. By Lemma 11.1b, it easily follows that we can find an $\alpha < \lambda$, a tuple $\mathbf{r}$ from X of the form $\mathbf{r}'\mathbf{r}''$ where all entries of $\mathbf{r}''$ are from $X(\alpha)$ and, for each $i = 1, 2, \dots, m$, a regular formula $\psi_i(y, \mathbf{x}'')$ from $\mathbf{L}_\alpha(NZFC)$ with $\mathbf{x}''$ having the arity of $\mathbf{r}''$ such that

$$p_i' = \{y \in X'(\alpha) : \mathcal{X}'(\alpha) \models \psi_i(y, \mathbf{r}'')\}.$$

Let $\overline{\{\mathbf{r}\}}$ be named by the $\mathbf{L}_\lambda(NZFC)$ predicate symbol $\mathcal{R}$. Then in $\mathcal{X}'$ the following $\mathbf{L}_\lambda(NZFC)$-sentence (note: *non*regular) holds concerning $\mathbf{p}$:

"There exists tuple $\mathbf{x}'\mathbf{x}''$ of elements from X such that
$\mathcal{R}(\mathbf{x}'\mathbf{x}'')$ and $\mathbf{x}' = \mathbf{r}'$ and for all $i = 1, 2, \dots, m$
$p_i' = \{y \in X'(\alpha) : \mathcal{X}'(\alpha) \models \psi_i(y, \mathbf{x}'')\}$ holds."

By matching types, the analogous sentence must hold concerning $\mathbf{q}$ and we get a tuple $\mathbf{s} = \mathbf{s}'\mathbf{s}''$ simply corresponding to $\mathbf{r}$ for which, writing $\mathbf{q}' = q_1', q_2', \dots, q_m'$, we get in each case that

$$q_i' = \{y \in Y'(\alpha) : \mathcal{Y}'(\alpha) \models \psi_i(y, \mathbf{s}'')\}.$$

Thus, $\mathbf{p}$ and $\mathbf{q}$ are corresponding tuples.

I now argue the converse. We can suppose the lemma is true for all orders $< \lambda$, but is false at order λ. Pick the simplest formula $\varphi(\mathbf{x})$ possible from $\mathbf{L}_\lambda(NZFC)$ such that $\mathbf{x}$ is m-ary and there exist corresponding $\mathbf{p} \in X'^m$ and $\mathbf{q} \in Y'^m$ for which

$$\mathcal{X}' \models \varphi(\mathbf{p}) \text{ and } \mathcal{Y}' \models \neg\varphi(\mathbf{q})$$

are both true. I shall refer to this as the *presumed counterexample*.

I first argue that since $\varphi(\mathbf{x})$ is of minimal complexity, it must indeed be atomic. Suppose otherwise. Since $\varphi(\mathbf{x})$ cannot be a boolean combination of other formulas and $\forall$ can be written as $\neg\exists\neg$, we can assume $\varphi(\mathbf{x})$ is some $(\exists y)\psi(y, \mathbf{x})$. From

$$\mathcal{X}' \models \varphi(\mathbf{p})$$

we get some $r \in X'$ such that

$$\mathcal{X}' \models \psi(r, \mathbf{p}).$$

By Lemma 11.1e, we can find $s \in Y'$ such that $r\mathbf{p}$ and $s\mathbf{q}$ also correspond. By induction on formulas, we get

$$\mathcal{Y}' \models \psi(s, \mathbf{q}) \text{ and hence } \mathcal{Y}' \models \varphi(\mathbf{q}),$$

which contradicts the presumed counterexample.

Thus $\varphi(\mathbf{x})$ must be atomic and in particular, we must have $\lambda = $ successor λ_0+1. Let simply corresponding $\mathbf{r} \in X^n$ and $\mathbf{s} \in Y^n$ generate the correspondence of $\mathbf{p} \in X'^m$ and $\mathbf{q} \in Y'^m$. Call an index i among $1, 2, \ldots, m$ *simple* if the ith entries p_i, q_i from $\mathbf{p}$ and $\mathbf{q}$ coincide with matching entries from $\mathbf{r}$ and $\mathbf{s}$. For nonsimple index i we can choose regular formula $\psi_i(y, \mathbf{x})$ from $\mathbf{L}_{\lambda_0}(NZFC)$ with $\mathbf{x}$ n-ary so that

$$p_i = \{r \in X'(\lambda_0) : \mathcal{X}'(\lambda_0) \models \psi_i(r, \mathbf{r})\}$$
$$q_i = \{r \in Y'(\lambda_0) : \mathcal{Y}'(\lambda_0) \models \psi_i(r, \mathbf{s})\}.$$

Case 1: $\varphi(\mathbf{x})$ is $x_1 = x_2$. We can assume $m = 2$. Since $\mathbf{r}$ and $\mathbf{s}$ correspond simply, at least some index i must be nonsimple. If all the indices are nonsimple then

$$\mathcal{X}' \models p_1 = p_2 \Leftrightarrow$$
$$\mathcal{X}'(\lambda_0) \models (\forall y)[\psi_1(y, \mathbf{r}) \leftrightarrow \psi_2(y, \mathbf{r})] \Leftrightarrow \text{ (induction)}$$
$$\mathcal{Y}'(\lambda_0) \models (\forall y)[\psi_1(y, \mathbf{s}) \leftrightarrow \psi_2(y, \mathbf{s})] \Leftrightarrow$$
$$\mathcal{Y}' \models q_1 = q_2,$$

which contradicts the presumed counterexample. Thus, it must be that (say) index 1 is simple and 2 is not. Since we have assumed $\mathcal{X}' \models p_1 = p_2$, we must have

$$p_1 \subseteq X(\lambda_0) \text{ (since 1 is simple) and}$$
$$\mathcal{X}'(\lambda_0) \models (\forall y)[\psi_2(y, \mathbf{r}) \rightarrow int(y)] \text{ (since } p_1 = p_2)$$

and hence (induction), that $q_1, q_2 \subseteq Y(\lambda_0)$. From the assumption $\mathcal{Y}' \models q_1 \neq q_2$ we must have some $s \in Y(\lambda_0)$ such that

$$\mathcal{Y}' \models \neg(s \in q_1 \leftrightarrow s \in q_2).$$

Using "ω-homogeneity" pick $r \in X(\lambda_0)$ so that $r\mathbf{r} \in X^{n+1}$ and $s\mathbf{s} \in Y^{n+1}$ also simply correspond. Then $rp_1 \in X^2$ and $sq_1 \in Y^2$ simply correspond so that

$$\mathcal{Y}' \models s \in q_1 \Leftrightarrow \mathcal{X}' \models r \in p_1.$$

We also have that

$$\mathcal{Y}' \models s \in q_2 \Leftrightarrow \mathcal{Y}'(\lambda_0) \models \psi_2(s, \mathbf{s})$$
$$\Leftrightarrow \text{ (induction) } \mathcal{X}'(\lambda_0) \models \psi_2(r, \mathbf{r})$$
$$\Leftrightarrow \mathcal{X}' \models r \in p_2,$$

which implies

$$\mathcal{X}' \models \neg(r \in p_1 \leftrightarrow r \in p_2),$$

which contradicts the presumed counterexample. Thus, case 1 is eliminated.

Case 2: $\varphi(\mathbf{x})$ is any of $int(x_1)$, $st(x_1)$, $\mathcal{R}(\mathbf{x})$, or for $\alpha < \lambda$, one of $\chi_\alpha(x_1)$ or $\chi'_\alpha(x_1)$. By Lemma 11.1e, we can find tuples $\mathbf{r}_0$ and $\mathbf{s}_0$ from X and Y so that $\mathbf{r}_0\mathbf{r}$ and $\mathbf{s}_0\mathbf{s}$ are simply corresponding and generate a correspondence of $\mathbf{p}\mathbf{p}$ to some $\mathbf{q}_0\mathbf{q}$. It can be arranged for any given matching entries p_0, q_0 from $\mathbf{p}$, $\mathbf{q}_0$ that if $p_0 \in X$, then $q_0 \in Y$ and they both coincide with some matching entries of $\mathbf{r}_0$, $\mathbf{s}_0$. Thus,

$$\mathcal{X}' \models \varphi(\mathbf{p}) \Rightarrow \mathcal{Y}' \models \varphi(\mathbf{q}_0)$$

when $\varphi(\mathbf{x})$ is any of $int(x_1)$, $st(x_1)$, $\mathcal{R}(\mathbf{x})$, or $\chi_\alpha(x_1)$ for $\alpha < \lambda$. If instead $p_0 \in X'(\alpha)$ for some $\alpha < \lambda$, then it can alternatively be arranged that $q_0 \in Y'(\alpha)$ as well. Thus, for all cases, we have

$$\mathcal{X}' \models \varphi(\mathbf{p}) \Rightarrow \mathcal{Y}' \models \varphi(\mathbf{q}_0).$$

But by case 1, since $\mathbf{p} = \mathbf{p}$ we have $\mathbf{q}_0 = \mathbf{q}$, and this contradicts the presumed counterexample.

Case 3: $\varphi(x)$ is $x_1 \in x_2$. Again, we may assume $m = 2$. Again, some index needs to be nonsimple. If the index 2 were simple, then $p_1 \in p_2 \subseteq X(\lambda_0)$ and case 2 would allow index 1 (and hence both) to be simple, so 2 must be a nonsimple index. From the assumption $\mathcal{X}' \models p_1 \in p_2$ we get that $p_1 \in p_2 \subseteq X'(\lambda_0)$ so that (case 2) we also have $q_1 \in Y'(\lambda_0)$. But then,

$$
\begin{aligned}
\mathcal{X}' \models p_1 \in p_2 &\Leftrightarrow \mathcal{X}'(\lambda_0) \models \psi_2(p_1, \mathbf{r}) \\
&\Leftrightarrow (\text{induction}) \ \mathcal{Y}'(\lambda_0) \models \psi_2(q_1, \mathbf{s}) \\
&\Leftrightarrow \mathcal{Y}' \models q_1 \in q_2
\end{aligned}
$$

which also contradicts the presumed counterexample.

Thus, all cases are eliminated and the lemma is proved. $\square$

The main value of the preceeding lemma lies in the fact that its criterion for similarity of $\mathbf{L}_\lambda(NZFC)$-types is "local" and hence, quite flexible. The next lemma illustrates the point.

LEMMA 11.3. *Assume $i : X \to Y$ is a morphism of $NZFC$-domains where X has order λ. Then i uniquely extends to a map $i' : X' \to Y'(\lambda) \subseteq Y'$ which is an elementary embedding $\mathcal{X}' \to \mathcal{Y}'(\lambda)$ of $\mathbf{L}_\lambda(NZFC)$-structures. This elementary embedding has the following special property: to each tuple $\mathbf{q} \in Y'(\lambda)^n$ there exists tuple $\mathbf{p} \in X'^n$ such that $\mathbf{q}$ and $i'(\mathbf{p})$ are of the same $\mathbf{Y}'(\lambda)$-types and are of the same $\mathbf{Y}'$-types as well.*

PROOF. Let $\mathbf{p} \in X'^n$ be any tuple. Use Lemma 11.1c to find appropriate $\mathbf{r} \in X^m$ and formula $\psi(\mathbf{x}, \mathbf{z})$ from $\mathbf{L}_\lambda(NZFC)$ such that

$$\mathcal{X}'(\lambda) \models (\forall \mathbf{x})[\mathbf{x} = \mathbf{p} \leftrightarrow \psi(\mathbf{x}, \mathbf{r})].$$

Then any $i' : X' \to Y'(\lambda) \subseteq Y'$ extending $i : X \to Y$ which gives an elementary embedding $\mathcal{X}' \to \mathcal{Y}'(\lambda)$ of the $\mathbf{L}_\lambda(NZFC)$-structures must also satisfy

$$\mathcal{Y}'(\lambda) \models (\forall \mathbf{x})[\mathbf{x} = i'(\mathbf{p}) \leftrightarrow \psi(\mathbf{x}, i(\mathbf{r}))].$$

Thus, i' is unique if it exists. Examining, first, the construction of the formula $\psi(\mathbf{x}, \mathbf{z})$ in the proof of Lemma 11.1c and, second, the construction in the proof of Lemma 11.1d of the tuple $\mathbf{q}$ from Y' which corresponds to $\mathbf{p}$, it is clear that there exists a (necessarily unique) tuple $\mathbf{q} \in Y'^n$ for which

$$\mathcal{Y}'(\lambda) \models (\forall \mathbf{x})[\mathbf{x} = \mathbf{q} \leftrightarrow \psi(\mathbf{x}, i(\mathbf{r}))].$$

By Lemma 11.2, this $\mathbf{q}$ is canonical and doesn't depend on the choice of $\mathbf{r}$ or $\psi(\mathbf{x}, \mathbf{z})$. We thus arrive at a canonically defined map $i' : X' \to Y'(\lambda)$ extending i for which $i'(\mathbf{p}) =$ this $\mathbf{q}$. That $\mathbf{p}$ and $i'(\mathbf{p})$ correspond in all cases shows they have matching $\mathbf{L}_\lambda(NZFC)$-types and that i' induces an elementary embedding $\mathcal{X}' \to \mathcal{Y}'(\lambda)$ of $\mathbf{L}_\lambda(NZFC)$-structures.

Now let $\mathbf{q}$ be an arbitrary tuple from $Y'(\lambda)$. Use Lemma 11.1d to find tuple $\mathbf{p}$ from X' to which $\mathbf{q}$ corresponds. But then both $\mathbf{q}$ and $i'(\mathbf{p})$ correspond to $\mathbf{p}$ and thus, have the same type in $\mathcal{Y}'(\lambda)$. Since $\mathbf{q}$ and $i'(\mathbf{p})$ correspond in $Y'(\lambda)$, by Lemma 11.1a, they continue to correspond in Y' and hence have the same type in $\mathcal{Y}'$. Since $\mathcal{Y}'(\lambda)$-types refine $\mathbf{Y}'(\lambda)$-types and $\mathcal{Y}'$-types refine $\mathbf{Y}'$-typles we are done. $\square$

—————CONCLUDING NS₁—————

We are finally able to complete the proof of correctness of our proposed model of NS_1. We assume that the $NZFC$-domain X is of order $\lambda = On_0$ and is the direct limit of a definable directed system

$$\{i_{\alpha\beta} : X[\alpha] \to X[\beta]\}_{\alpha < \beta \in On_0}$$

of small $NZFC$-domains such that for each cardinal $\kappa \in On_0$, $X[\alpha + 1]$ is κ-saturated and compact for all sufficiently large $\alpha \in On_0$.

I now argue that $\mathbf{X}' = (X', \in |_{X'}, X, {}^\circ X)$ with $X' = X'(\lambda)$ is a model of NS_1. Due to our previous work, it is left only to verify that $\mathbf{X}'$ is definable, satisfies separation and satisfies collection.

Proof of definability: Since each $X[\alpha]$ is definable and small, its $\mathbf{L}(NZFC)$-structure $\mathbf{X}'[\alpha]$, being canonically and inductively formed from $X[\alpha]$, is also definable and small. These definitions can be done uniformly so the very sequence of $L(NZFC)$-structures

$$\{\mathbf{X}'[\alpha]\}_{\alpha \in On_0}$$

is definable. The lifting of the maps $i_{\alpha\beta} : X[\alpha] \to X[\beta]$ to maps $i'_{\alpha\beta} : X'[\alpha] \to X'[\beta]$ as described in Lemma 11.3 is also canonical and thus, the entire system of lifted maps

$$\{i'_{\alpha\beta} : X'[\alpha] \to X'[\beta]\}_{\alpha < \beta \in On_0}$$

is definable. With these definitions in place, an isomorphic copy of $\mathbf{X}'$ as a direct limit of the $\mathbf{X}'[\alpha]$ can now be constructed within V_0 and this shows that $\mathbf{X}'$ is definable.

Before taking up proofs of separation and collection, we need additional notation and one key observation.

Notation: For each $\alpha \in On_0$ let $i_\alpha : X[\alpha] \to X$ be the canonical morphism into the direct limit X. Without loss of generality, we can assume the order of each $X[\alpha]$ is α. By Lemma 11.3, the map i_α lifts to an elementary embedding

$i'_\alpha : \mathcal{X}'[\alpha] \to \mathcal{X}'(\alpha)$. I shall abbreviate $i'_\alpha[X'[\alpha]] \subseteq X'(\alpha)$ and $i_\alpha[X[\alpha]] \subseteq X(\alpha)$ as $X'\{\alpha\}$ and $X\{\alpha\}$, respectively, and put

$$\mathbf{X}'\{\alpha\} = (X'\{\alpha\}, \in |_{X'\{\alpha\}}, X\{\alpha\}, {}^\circ X(\alpha)).$$

A key observation: The $NZFC$-structures $\mathbf{X}'\{\alpha\}$ for $\alpha \in On_0$ form a homomorphically increasing continuous system of *small* substructures of $\mathbf{X}'$ whose limit is $\mathbf{X}'$ which is *large*. By V_0-collection transfered to X', an extended reflection principle on X' holds and thus, for each formula $\varphi(\mathbf{x})$ of $\mathbf{L}(NZFC)$ there will exist a closed unbounded class of $\alpha \in On_0$ for which $\mathbf{X}'\{\alpha\}$ reflects $\varphi(\mathbf{x})$ for $\mathbf{X}'$. It turns out for these same α that $\mathbf{X}'(\alpha)$ will *also* reflect $\varphi(\mathbf{x})$ for $\mathbf{X}'$. Indeed, suppose $\mathbf{x}$ is n-ary and choose $\mathbf{p} \in X'(\alpha)^n$. By Lemma 11.3, we can choose $\mathbf{p}_0 \in X'\{\alpha\}^n$ having the same type in $\mathbf{X}'\{\alpha\}$ that $\mathbf{p}$ has in $\mathbf{X}'(\alpha)$. Also, by this lemma, these $\mathbf{p}_0$ and $\mathbf{p}$ continue to have the same type when considered as lying in $\mathbf{X}'$. We get then that

$$\mathbf{X}'(\alpha) \models \varphi(\mathbf{p}) \Leftrightarrow \text{ (same type)}$$
$$\mathbf{X}'\{\alpha\} \models \varphi(\mathbf{p}_0) \Leftrightarrow \text{ (reflection)}$$
$$\mathbf{X}' \models \varphi(\mathbf{p}_0) \Leftrightarrow \text{ (same type)}$$
$$\mathbf{X}' \models \varphi(\mathbf{p})$$

showing that $\mathbf{X}'(\alpha)$ also reflects $\varphi(\mathbf{x})$ for $\mathbf{X}'$.

Proof of separation: Let $p \in X'$ and let $\varphi(x, \mathbf{z})$ be a formula from $\mathbf{L}(NZFC)$ where $\mathbf{z}$ is n-ary. Fix any tuple $\mathbf{r}$ from X'^n. Pick $\alpha \in On_0$ such that $\mathbf{X}'(\alpha)$ reflects $\varphi(x, \mathbf{z})$ for $\mathbf{X}'$ and $X'(\alpha)$ contains p and the entries of $\mathbf{r}$. By construction of $X'(\alpha + 1)$, there exists $s \in X'(\alpha + 1)$ such that

$$s = \{t \in X'(\alpha) : \mathbf{X}'(\alpha) \models t \in p \,\&\, \varphi(t, \mathbf{r})\}.$$

But then by reflection, the fact that $X'(\alpha) \subseteq X'$ is transitive and the fact that $p \subseteq X'(\alpha)$, we have

$$s = \{t \in X' : \mathbf{X}' \models t \in p \,\&\, \varphi(t, \mathbf{r})\}$$

and hence, that

$$\mathbf{X}' \models (\exists y)(\forall x)[x \in y \leftrightarrow x \in p \,\&\, \varphi(x, \mathbf{r})].$$

This shows the axiom of separation holds.

Proof of collection: Let $\varphi(u, v, \mathbf{z})$ be a formula from $\mathbf{L}(NZFC)$ where $\mathbf{z}$ is n-ary. Fix $p \in X'$ and n-ary $\mathbf{r}$ with entries from X'. The axiom of collection requires that there exist $p' \in X'$ such that

$$\mathbf{X}' \models (\forall u \in p)[(\exists v)\varphi(u, v, \mathbf{r}) \to (\exists v \in p')\varphi(u, v, \mathbf{r})].$$

Pick $\alpha \in On_0$ such that $X'(\alpha)$ contains p and the entries of $\mathbf{r}$ and such that $\mathbf{X}'(\alpha)$ reflects *both* $\varphi(u, v, \mathbf{r})$ and $(\exists v)\varphi(u, v, \mathbf{r})$ for $\mathbf{X}'$. Set $p' = X'(\alpha)$. Then for any $s \in p$,

$$\mathbf{X}' \models (\exists v)\varphi(s, v, \mathbf{r}) \Rightarrow \text{(reflection)}$$
$$\mathbf{X}'(\alpha) \models (\exists v)\varphi(s, v, \mathbf{r}) \Rightarrow$$
$$\text{for some } q \in p', \mathbf{X}'(\alpha) \models \varphi(s, q, \mathbf{r}) \text{ holds } \Rightarrow \text{(reflection)}$$
$$\text{for some } q \in p', \mathbf{X}' \models \varphi(s, q, \mathbf{r}) \text{ holds } \Rightarrow \mathbf{X}' \models (\exists v \in p')\varphi(s, v, \mathbf{r}),$$

which shows that $\mathbf{X}'$ satisfies collection.

The proof that $\mathbf{X}'$ is a valid model of NS_1 is now complete. $\square$

Discussion: I would like to end this chapter with a few historical remarks and acknowledgements. A number of ideas used here were borrowed from Hrbáček's important paper [**11**]. The uniform strategy of constructing models of "non"-standard set theories within a conservative extension of ZFC (for us: AST) whose standard parts are isomorphic to a small inner model of ZFC (for us: U_0) is essentially taken from [**11**]. In the case of modeling NS_1, I have explicitly used Hrbáček's technique of constructing a model which is *definable* within the inner ZFC model and exploiting the resulting ability to transport collection-style arguments outward. Although the present model of NS_1 is roughly parallel to Hrbáček's, it is simpler in that it avoids use of constructible sets based on infinitary languages. Nevertheless, the present proof of the model's correctness also has rough parallels to Hrbáček's. Specifically, the elementary embedding mentioned in Lemma 11.3 has similarities to Hrbáček's Lemma 6 in [**11**]. I think, however, one can say that the type analysis achieved in the present Lemma 11.2 represents a new and deeper insight as to why the eventual NS_1 models do work. The technique of augmenting a structure $\mathbf{X}'$ to a structure $\mathcal{X}'$ to achieve a refined and more convenient system of types appears to me to be new and I would like to call the model theorist's attention to it as it may have further applications.

CHAPTER 12

Critical Review With Proposal: *EST*

The "non"standard set theories of Nelson, Kawai and Hrbáček investigated in the preceding chapters represent various attempts to fulfill the dream of a global vehicle for "non"standard mathematical practice suitable for use by ordinary mathematicians. More recently, in [**9**], Fletcher discussed the short comings of these theories and offered his own improvement, which he called *Stratified* Nonstandard Set Theory, or *SNST* for short.

I shall summarize Fletcher's criticisms (with which I agree) and then argue that his own *SNST* has problems as well. Pressing further, I will analyze not only the constraints (discovered chiefly by Hrbáček) for any such vehicle, but also consider the philosophical tendencies which the various attempts at it seem to suggest. These philosophical tendencies have a discernable direction along which the previous theories so far only partly progress. I conclude this chapter by describing a prototype of a "non"standard set theory which follows these philosophical tendencies to their logical conclusion. Called *EST*, for **E**nlargement **S**et **T**heory, it is intended to illustrate what the "ultimate" vehicle for "non"standard mathematical practice might look like.

The following shopping list of desired features has emerged from the numerous attempts at finding "the" global vehicle for "non"standard mathematical practice:

1) There should exist an internal universe elementarily extending the standard universe in a manner that is "sufficiently" saturated.

2) An external universe extending the internal universe should also exist which satisfies full ZF^-C, and in which "standardization" of external sets is possible.

In the second feature "full" means that the ZF^-C axiom schemata of separation and collection would remain valid when taken with respect to an appropriately extended language that distinguished standard, internal and external sets. Also, the demand for "full ZF^-C" rather than "full ZFC" reflects the fact that any saturation in the internal universe forces a failure of the axiom of regularity in the external universe.

Historically the first to be published, Nelson's *IST* [**18**] made no attempt at an external universe, although it did allow for standardization of definable subclasses of standard sets. The internal universe in *IST* got "sufficient" saturation by being globally ω-saturated.

Independently, Hrbáček [**11**] next outlined his theories NS_1 and NS_2 and uncovered the first major constraints for any global vehicle for "non"standard mathematical practice. He discovered that in a rather strong sense a naive interpretation of features (1) and (2) is simply impossible. For "sufficient" saturation he demanded, in essence, that arbitrarily large standard sets be arbitrarily saturated (with respect to cardinals within the theory). He also imposed arbitrary standardization for external sets. Given this, he proved that the external universe must not only fail regularity, but also, fail *either* collection *or both* power sets and choice. His theories NS_1 and NS_2 were designed to be best possible compromises, given these facts: NS_1 allowed external collection, while NS_2 allowed both external power sets and choice.

And there matters sat for a while. It wasn't clear where to place the blame for these constraints, but the demand for "sufficient" saturation in the internal universe seemed a plausible candidate.

Subsequently, Kawai shed further light on the matter. In [**14**] he proposed NST as a direct extension of Nelson's *IST*, which now included a true universe of externals. He allowed increased saturation for the internal universe but retained *IST*'s restricted standardization. The internal universe is now card$(S)^+$-saturated and any external set can be standardized, provided it is already contained in a standard set. With this, Kawai was able to have an external universe satisfying *full ZF^-C*. It even satisfies a modified axiom of regularity which says, in effect, that the external universe is regular over the internal universe.

Thus, blame for the Hrbáček constraints could now seem to shift back to the demand for unlimited standardization.

But as pointed out by Fletcher [**9**], even Kawai's NST is too confusing to be useful for ordinary practitioners: Unlike ZFC, external sets in NST can be of unlimited size and there is a perpetual uncertainty as to when one of these can be standardized. Fletcher's contribution is to return to the demand in feature (1) for "sufficient" saturation and to point out that from the practitioner's point of view, the previous theories had involved overkill. Typically, the practitioner works in a local portion of the set universe and does not need unbounded saturation at any one time. In Fletcher's $SNST$, the internal universe I is a direct limit of internal "stages" I_κ (for standard cardinal κ), each of which is a κ-saturated elementary extension of the standard universe S. At each "stage" κ, a full ZF^-C universe E_κ of external sets regular over I_κ is allowed, and the external universe E is the grand union of these. Arbitrary standardization of external sets is permitted. By working in a particular internal I_κ and external E_κ — for high enough κ — Fletcher allows sufficient saturation along with externals and their standardization to suit the needs of the practitioner at any one time. In his theory $SNST$, the final internal universe I and the final external universe E are of no particular interest: in fact, quantifiers are only used over specific I_κ's and E_κ's.

And this is the source of an immediate criticism of Fletcher's $SNST$: notationally it is a thicket. The theory involves heavy use of quantifiers of the form (Q^α) and $(Q^{ext,\alpha})$ and really represents a throwback to the days of Whitehead and Russell's (see [**25**]) ramified theory of sets.

In addition, Hrbáček pointed out to me a deeper criticism which applies not only to Fletcher's $SNST$, but to all the "non"standard set theories under review here. The global vehicle one sought was implicitly meant to *improve* on Robinson's enlargements and not give up on any of their flexibility. One flexibility of Robinson's enlargements was that they could be "layered": the external sets of one enlargement could be the standard sets in another. For Robinson, the notions of standard, internal, and external sets were always relative to the enlargement of the moment. By contrast, in each of the "non"standard set theories considered here, these notions are statically fixed for all time. Although there has apparently been no use so far of a "layered" "non"standard argument by practitioners, I am convinced that an optimum global vehicle for "non"standard mathematics should include the feature

    3) The notions of standard, internal, and external sets should be sufficiently relativized to allow for arbitrary "layering" of "non"standard arguments.

The foregoing completes the shopping list I propose for an "ultimate" global vehicle for "non"standard mathematical practice. Philosophically, the very first attempts to achieve this vehicle involved a jump from the ordinary mathematician's sense of ontology: Instead of a uniform sea of sets, the universe now consisted of sets *of different types* (e.g., standard, internal, external). Indeed, this enriched ontology provided some of the very leverage of "non"standard mathematics. However, a jump in ontology is a jump in fundamentals and this perhaps explains why each of the "non"standard set theories we have considered stopped well short of feature (3) on the shopping list above. I agree with Fletcher that a variable but bounded saturation is a proper compromise to the constraints discovered by Hrbáček. What is further needed is a relativization of the notions of standardness, internalness, and externalness to accommodate feature (3).

For this purpose, I propose a simple yet nontrivial modification of Fletcher's $SNST$: For each cardinal κ, the internal universe I_κ shall be a κ-saturated elementary extension, not of S, but of E_κ. Thus, each external set will get to play the role of a standard set in a further enlargement of things.

This new theory is to be called EST, for **E**nlargement **S**et **T**heory. It remains to be axiomatized. In doing so we need to avoid Fletcher's notational problems. Essentially, we need to do to Fletcher's $SNST$ what Zermelo did to Whitehead and Russell's ramified theory of sets. We shall let the language of EST be that of *BG* set theory, that is, it is to have variables ranging over classes and have as sole nonlogical constant the binary predicate symbol $\in$, indicating membership. As done previously, we use $A, B, C \ldots$ as informal class variables. Sets are defined as usual as elements of classes, and we again use $x, y, z \ldots$ as informal set variables.

Our intended interpretation for *EST* is of a cosmos of sets undergoing perpetual expansion. The expansion will be twofold: First, there will be *external* expansion wherein standard set theoretic operations (power sets, unions, cartesian products, collection, etc.) are used to form new sets from old. Secondly, there will be *internal* expansion wherein any set x, unless finite, will continually pick up new elements through the "non"standard process of saturation.

In *EST* we shall use classes which are not sets to serve as frozen still shots of how the totality of all sets looks at various stages and points of view. We call a class B *static* if it is a subclass of some class A which is not a set. In *EST*, this will typically amount to saying that either B is not a set or if it is, then it is finite. A static class B which is a subclass of a set will be called *local*[1]. Whether local or not, a static class A will "see" a set x only if $x \in A$, and what A will "see" of a class B will only be its "trace" on A, namely, $B \cap A$. The most useful still shots of the unfolding totality of *EST* sets will be those static classes A which perceive an "ordinary" mathematician's universe. We will call such a class A a *universe*.

To properly spell this out, we shall need some notation. For the rest of this chapter we shall consider that a formula φ from *EST*'s language shall come with an (informal) understanding concerning which of its variables (free or not) are set variables and which are class variables. We will write formulas with informal set and class variables to indicate this understanding. For a formula $\varphi(\mathbf{z}, \mathbf{D})$ in the language of *EST* with free set variables among the $\mathbf{z} = z_1, \ldots z_n$ and free class variables among the $\mathbf{D} = D_1, \ldots D_m$, I will write

$$A \models \varphi(\mathbf{z}, \mathbf{D})$$

to mean the formula in *EST*'s language which expresses

$$z_1 \in A \,\&\, \ldots \,\&\, z_n \in A \,\&\, D_1 \subseteq A \,\&\, \ldots \,\&\, D_m \subseteq A \,\&\, {}^A\varphi(\mathbf{z}, \mathbf{D})$$

where ${}^A\varphi(\mathbf{z}, \mathbf{D})$ is obtained from $\varphi(\mathbf{z}, \mathbf{D})$ by rewriting set quantifiers (Qx) as $(Qx \in A)$ and class quantifiers (QB) as $(QB \subseteq A)$.

A **universe** shall be any *non*local static clsss with the following properties:

•) As a class, A "sees" a model of $BG^- C^+$, slightly modified. This modification, which we write as $BG^- C^+_{weak\,ext}$, is obtained by weakening the axiom of extentionality to read as the conjunction of

$$(\forall z)[z \in B \leftrightarrow z \in C] \rightarrow B = C \text{ and}$$
$$(\forall z)[z \in x \leftrightarrow z \in y] \rightarrow x = y.$$

---

[1]In [**24**] these classes were called *semisets* .

By the conventions just adopted, this means from A's point of view that sets are extensional and static classes are extensional but that a set x and a static class B are allowed to have a common extension (as far as A is concerned) and yet, still retain their separate identities. When this happens, we will express matters by saying that B is A's *static copy* of the set x. Comprehension axioms in $BG^-C^+_{weak\,ext}$ shall be understood to assert the existence of static subclasses of A. Of course those axioms of $BG^-C^+_{weak\,ext}$ which assert the existence of sets (e.g., $\{x,y\}$, $\bigcup x$, etc.) must be carefully expressed to assert the existence of a set whose extension (from A's point of view) is that of the appropriate (informal) class. For example, one uses

$$A \models (\forall B)(\forall x)(\exists y)(\forall z)[z \in y \leftrightarrow z \in x \ \& \ z \in B]$$

to express the fact that A satisfies the axiom of separation.

•) A must also satisfy full separation with respect to arbitrary classes, i.e., for each (actual) class B the informal class $B \cap A$ should actually exist. In symbols, this is

$$(\forall B)(\exists C)(\forall x)[x \in C \leftrightarrow x \in B \ \& \ x \in A].$$

In particular, this forces A to have a static copy of each of the sets it sees.

•) To eliminate certain pathologies, A should also be able to correctly detect when a static subclass is local or when it is a set. Expressed in symbols, we want

$$(\forall B \subseteq A)[B \text{ local} \ \rightarrow \ \text{``}A \models B \text{ local''}]$$
$$(\forall x)[x \subseteq A \rightarrow x \in A].$$

Since A satisfies the axiom of separation (for sets), this implies that any local static subclass B of A is in fact the static A-copy of some set seen by A. In particular, for any set x, A sees a set $°x$ whose static A-copy is $x \cap A$.

•) Finally, to be a universe the class A should reflect restricted formulas for any universe $A \subseteq A'$ extending it. We need to remove the obvious circularity of this assertion and also find a way to express its content within *EST*'s language. Since A and A' both model ZF^-, the remarks at the end of Chapter 8 show how to express A's reflection for A' of arbitrary restricted formulas as a single *EST*-formula, namely, assert that the Gödel operations are preserved. This still leaves circularity. Luckily, we may remove that by asserting that the Gödel operations are preserved in the extension of A to the cosmology of *all EST* sets. Namely, for each Gödel operation we choose a standard restricted formula $\theta(\mathbf{z}, y)$ defining it and require of A that

$$(\forall \mathbf{z}, y \in A)[\text{``}A \models \theta(\mathbf{z}, y)\text{''} \leftrightarrow \theta(\mathbf{z}, y)].$$

The conjunction of these (finitely many) formulas completes the requirements for A to be a universe. These imply for any extension $A \subseteq A'$ of universes (as now defined) and restricted formula $\theta(\mathbf{z}, y)$ defining a Gödel operation, that

$$A \text{ reflects } \theta(\mathbf{z}, y) \text{ for } A'.$$

Thus, the Gödel operations are preserved in the extension $A \subseteq A'$ and therefore, as desired, A reflects arbitrary restricted formulas for A'.

A number of set theoretic constructions performable in a universe A will now appear to be *absolute*. For example, to any sets x, $y \in A$, there will correspond further sets in A which (for A) play the role of $\{x, y\}$, $< x, y >$, $x \times y$, $x - y$, $x \cap y$, $\bigcup x$, etc., and these sets will continue to play these roles when viewed in any larger universe A' extending A [2]. Similarly, A and A' will agree on who the emptyset $\emptyset$ is. For any restricted formula $\varphi(z, \mathbf{y})$ the same can be said for the operation that forms (relative to x and $\mathbf{y}$) the set $\{z \in x : \varphi(z, \mathbf{y})\}$. An important counterexample to this phenomenon, however, is the power set operation: Although universes $A \subseteq A'$ may both see a set x, unless x is finite they typically will disagree on who its power set is.

Before listing the actual axioms of *EST*, I need to develop a clearer picture of its universes and their extensions. The reader should keep in mind that discussion which follows is *axiom free* :

Even at this point, an extension $A \subseteq A'$ of universes carries its own *transfer principle*.

PROPOSITION 12.1. [transfer principle for universes] *Let $A \subseteq A'$ be universes and to each static class $B \subseteq A$ associate the unique static class $\overline{B} \subseteq A'$ [3] for which*

$$A' \models \overline{B} = \bigcup \{x \in A : A \models x \subseteq B\} \,.$$

Then $\overline{A} \subseteq A'$ is the transitive closure (for A') of $A \subseteq A'$ and $\overline{A}$ amounts to an elementary extension of A in the following sense: To each formula $\varphi(\mathbf{z}, \mathbf{D})$ whose only quantified variables are set variables, it is the case that

$$(\forall \mathbf{D} \subseteq A)(\forall \mathbf{z} \in A)[A \models \varphi(\mathbf{z}, \mathbf{D}) \leftrightarrow \overline{A} \models \varphi(\mathbf{z}, \overline{\mathbf{D}})] \,.$$

PROOF. I first argue for each static class $B \subseteq A$ that $B = \overline{B} \cap A$. Since these are both static subclasses of A', it suffices to show that they have the same elements. If $p \in B$, then $p \in \{p\} \subseteq B$ and $\{p\} \in A$, so we have $p \in \overline{B}$, and since $B \subseteq A$, we have $p \in \overline{B} \cap A$. Conversely, if $p \in \overline{B} \cap A$, then for some $p' \in A$ we have $p \in p'$ and $p' \cap A \subseteq B$ and, since $p \in A$, it follows that $p \in B$. The foregoing now implies for any static subclasses B, $B' \subseteq A$ that $B = B'$ if and only if $\overline{B} = \overline{B}'$.

Next, let $\widehat{A} \subseteq A'$ be the static subclass such that

$$A' \models \widehat{A} = \text{ the transitive closure of } A \,.$$

---

[2]For this reason I shall informally and freely write $\{x, y\}$, $< x, y >$, etc. for the sets in a universe A or its extensions A' which play (for them) the appropriate roles. The universe A and its A''s shall be understood from the context of the discussion.

[3]The topological notation here is intentionally suggestive.

I argue that $\overline{A} = \widehat{A}$. Again, we only need show the classes have a common extension of sets. If $p \in \overline{A}$, there exists $p' \in A$ such that $p \in p'$ and (redundantly) $p' \cap A \subseteq A$, so by definition $p \in \widehat{A}$. Conversely, suppose $p \in \widehat{A}$. Then there exists $p' \in A$ such that $p \in p'$ and since trivially $p' \cap A \subseteq A$, we have $p \in \overline{A}$.

Finally, let $\varphi(\mathbf{z}, \mathbf{D})$ be any formula in *EST*'s language with free set variables among the $\mathbf{z} = z_1, \ldots z_n$ and free class variables among the $\mathbf{D} = D_1, \ldots, D_m$ and whose only quantified variables are set variables. Let $\mathbf{p} = p_1, \ldots, p_n \in A$ and $\mathbf{B} = B_1, \ldots, B_m \subseteq A$ be appropriate tuples of A-sets and A-static classes. I want to show that

$$A \models \varphi(\mathbf{p}, \mathbf{B}) \Leftrightarrow \overline{A} \models \varphi(\mathbf{p}, \overline{B}).$$

We can assume some simplifications: If any B from $\mathbf{B}$ is a set, then by definition $B \in A$ and restricted reflection on $x \subseteq B$ shows $\overline{B} = B$. Thus, any such B can be considered as belonging to the tuple $\mathbf{p}$ or, put otherwise, that all entries B from $\mathbf{B}$ can be assumed to be nonsets (hence with *non*empty extensions). Given this, any atomic subformula "$\delta \in \ldots$" in $\varphi(\mathbf{p}, \mathbf{B})$, where δ is not a set variable or one of the sets from $\mathbf{p}$, must be of the form "$B \in \ldots$" where B is from $\mathbf{B}$ and, being a nonset, such an atomic subformula can for our purposes be replace by (say) a "$p_1 \neq p_1$". Similarly, any atomic subformula "$\delta = \delta'$" where one of δ, δ' is a static class from $\mathbf{B}$ and the other is not can again be replaced by a "$p_1 \neq p_1$". If in "$\delta = \delta'$" both δ and δ' are static classses from $\mathbf{B}$, then from our previous remarks this can be replaced by a "$p_1 = p_1$" if it is true and a "$p_1 \neq p_1$" if otherwise. The end result of these simplifications is that we can assume that in $\varphi(\mathbf{p}, \mathbf{B})$ the only atomic subformulas involving some B from $\mathbf{B}$ are of the form "$\delta \in B$" where δ is either a set variable or a set p from $\mathbf{p}$. Let us consider the "tuples"[4]

$$\mathbf{A} = (A, B_1, \ldots B_m) \text{ and } \overline{\mathbf{A}} = (\overline{A}, \overline{B}_1, \ldots \overline{B}_m)$$

as large ($=$ nonlocal) model theoretic structures with a common language $\mathbf{L}$. We can now consider $\varphi(\mathbf{p}, \mathbf{B})$ and $\varphi(\mathbf{p}, \overline{B})$ as ordinary $\mathbf{L}$-sentences about $\mathbf{A}$ and $\overline{\mathbf{A}}$. What we are obviously required to show is that $\mathbf{A}$ has $\overline{\mathbf{A}}$ as an elementary extension. Since A' is a universe, there exists static $F \subseteq A'$ such that

$$A' \models \text{"}F \text{ is an increasing continuous } On\text{-indexed sequence}$$
$$\{W_\alpha\}_{\alpha \in On} \text{ of elements of } A \text{ such that } A = \bigcup_{\alpha \in On} W_\alpha \cap A\text{".}$$

Let $On_{A'} \subseteq A'$ be the static class for which $A' \models On_{A'} = \{x : x \text{ an ordinal}\}$ and for each $\alpha \in On_{A'}$ let

$$\mathbf{A}_\alpha = (W_\alpha \cap A, W_\alpha \cap B_1, \ldots, W_\alpha \cap B_m)$$
$$\overline{\mathbf{A}}_\alpha = (W_\alpha \cap \overline{A}, W_\alpha \cap \overline{B}_1, \ldots, W_\alpha \cap \overline{B}_m)$$

be the induced small ($=$ local) $\mathbf{L}$-structures. If we write $\varphi'(\mathbf{x})$ for the $\mathbf{L}$-formula of which $\varphi(\mathbf{p}, \mathbf{B})$ and $\varphi(\mathbf{p}, \overline{B})$ are both "instances", then since A' is a universe

---

[4]The nonexistence in *EST* for such "tuples" via conventional encodings is an inessential problem: pick alternative encodings, letting (say) $\mathbf{A} = \{m\} \times A \times B_1 \times \cdots \times B_m$

and the extended reflection principle holds in it, it follows that for some $\alpha \in On_{A'}$ we have

$$A' \models \text{``}\mathbf{A}_\alpha \text{ reflects } \varphi'(\mathbf{x}) \text{ for } \mathbf{A}\text{''} \ \& \ \text{``}\overline{\mathbf{A}}_\alpha \text{ reflects } \varphi'(\mathbf{x}) \text{ for } \overline{\mathbf{A}}\text{''}$$
$$\& \ p_1, \ldots, p_n \in W_\alpha \cap A.$$

For any local static classes $C \subseteq A$ and $C' \subseteq A'$, let $\widehat{C} \in A$ and $\widehat{C'}^\dagger \in A'$ be the *sets* for which

$$A \models (\forall x)[x \in C \leftrightarrow x \in \widehat{C}]$$
$$A' \models (\forall x)[x \in C' \leftrightarrow x \in \widehat{C'}^\dagger],$$

and put

$$\widehat{\mathbf{A}_\alpha} = (\widehat{W_\alpha \cap A}, \widehat{W_\alpha \cap B_1}, \ldots, \widehat{W_\alpha \cap B_m})$$
$$\widehat{\mathbf{A}_\alpha}^\dagger = (\widehat{W_\alpha \cap A}^\dagger, \widehat{W_\alpha \cap B_1}^\dagger, \ldots, \widehat{W_\alpha \cap B_m}^\dagger).$$

A key observation for us is that in *EST*'s language both

$$\widehat{\mathbf{A}_\alpha} \models \varphi'(\mathbf{p}) \text{ and } \widehat{\mathbf{A}_\alpha}^\dagger \models \varphi'(\mathbf{p})$$

are *restricted* formulas.

Another essential observation is for any element $q \in A$ and static class $C \subseteq A$, that the sets $\widehat{q \cap C}$ and $\widehat{q \cap \overline{C}}^\dagger$ *coincide*. To verify this, it suffices to show that

$$\widehat{q \cap C} \cap A' = \widehat{q \cap \overline{C}}.$$

Since

$$A \models \widehat{q \cap C} \subseteq q,$$

restricted reflection gives

$$A' \models \widehat{q \cap C} \subseteq q.$$

Also,

$$\widehat{q \cap C} \in A \text{ and } A \models \widehat{q \cap C} \subseteq C$$

give

$$A' \models \widehat{q \cap C} \subseteq \overline{C},$$

and thus,

$$A' \models \widehat{q \cap C} \subseteq q \cap \overline{C}.$$

Conversely, suppose $A' \models r' \in q \cap \overline{C}$ is the case. Then for some $r \in A$

$$A' \models r' \in r \ \& \ r \cap A \subseteq C,$$

and since $q \in A$, we can also assume (replace r with $r \cap q$) that

$$A' \models r \cap A \subseteq q \cap C \cap A,$$

which implies

$$A' \models r \cap A \subseteq \widehat{q \cap C} \cap A,$$

and (more restricted reflection) we get

$$A' \models r \subseteq \widehat{q \cap C}, \text{ hence } A' \models r' \in \widehat{q \cap C}.$$

This shows

$$A' \models q \cap \overline{C} \subseteq \widehat{q \cap C}$$

and the claim is establised.

Putting this all together, we have

$$A \models \varphi(\mathbf{p}, \mathbf{B}) \Leftrightarrow A' \models \text{``}\mathbf{A} \models \varphi'(\mathbf{p})\text{''} \Leftrightarrow \text{(reflection)}$$

$$A' \models \text{``}\mathbf{A}_\alpha \models \varphi'(\mathbf{p})\text{''} \Leftrightarrow A \models \text{``}\widehat{\mathbf{A}_\alpha} \models \varphi'(\mathbf{p})\text{''} \Leftrightarrow \text{(restricted reflection)}$$

$$A' \models \text{``}\widehat{\mathbf{A}_\alpha} \models \varphi'(\mathbf{p})\text{''} \Leftrightarrow (\widehat{\mathbf{A}_\alpha} \text{ is } \widehat{\mathbf{A}_\alpha}^\dagger \text{ !})$$

$$A' \models \text{``}\widehat{\mathbf{A}_\alpha}^\dagger \models \varphi'(\mathbf{p})\text{''} \Leftrightarrow A' \models \text{``}\overline{\mathbf{A}}_\alpha \models \varphi'(\mathbf{p})\text{''} \Leftrightarrow \text{(reflection)}$$

$$\overline{A} \models \varphi(\mathbf{p}, \overline{B}),$$

and the proof is complete. $\square$

We next need to pin down a notion of when an extension $A \subseteq A'$ of universes is κ-*saturated*, for cardinal number κ. Since the concept of a cardinal number is not absolute, a technical maneuver is called for. The following proposition is essential for this.

PROPOSITION 12.2. *Let $A \subseteq A'$ be universes and let U_A and $U_{A'}$ be the static classes of A-regular and A'-regular sets. Then*

$$A' \models \text{``}(U_A, \in |_{U_A}) \approx (U_{A'}, \in |_{U_{A'}}) \text{ and the isomorphism is unique''}.$$

PROOF. Clearly

$$A \models (U_A, \in |_{U_A}) \text{ is extensional,}$$

which gives

$$A' \models (U_A, \in |_{U_A}) \text{ is extensional,}$$

and since for any $x \in U_A$, $x \cap A = x \cap U_A$ is the A-static copy of of some A'-set x', we also get

$$A' \models (\forall x \in U_A)(\exists x')(\forall z)[z \in x' \leftrightarrow z \in |x|_{\in |_{U_A}}].$$

In other words, from A''s point of view, $\in |_{U_A}$-represented collections "are sets".

On the other hand,

$$A \models (U_A, \in |_{U_A}) \text{ is well-founded}$$

holds. Now suppose

$$A' \models (U_A, \in |_{U_A}) \text{ is not well-founded}$$

were the case. Then x', $f' \in A'$ would exist such that

$$A' \models x' \subseteq U_A \;\&\; x' \neq \emptyset \;\&\; f' : x' \to x' \;\&\; (\forall y \in x')[f'(y) \in y].$$

But both $x' \cap A$ and $f' \cap A'$ are local static subclasses of A (recall ordered pairs are "absolute") and hence, are static A-copies of sets x, $f \in A$. But then clearly,

$$A \models x \subseteq U_A \;\&\; x \neq \emptyset \;\&\; f : x \to x \;\&\; (\forall y \in x)[f(y) \in y],$$

which contradicts the fact that A sees $(U_A, \in |_{U_A})$ as well-founded.

Therefore,

$$A' \models (U_A, \in |_{U_A}) \text{ is well-founded}$$

and by the Mostowski Collapsing Lemma (*global* form), there exists a unique static class $F \subseteq A'$ such that

$$A' \models F : (U_A, \in |_{U_A}) \to (U_{A'}, \in |_{U_{A'}}) \text{ is a } (1-1) \text{ homomorphism for which}$$
$$(\forall x \in U_A)(\forall z')[z' \in F(x) \leftrightarrow (\exists z \in x)(z' = F(z))].$$

If F were *not* an isomorphism (i.e., not onto), then the A'-well-foundedness of $(U_{A'}, \in |_{U_{A'}})$ and A'-global choice would imply the existence of $x' \in U_{A'}$ such that

$$A' \models x' \subseteq F[U_A] \ \& \ x' \notin F[U_A].$$

But by the axiom of replacement and the fact that F is $(1-1)$, there exists set $x'_0 \in A'$ such that A''s static copy of x'_0 is $F^{-1}[x' \cap A'] \subseteq U_A$. Thus, $F^{-1}[x' \cap A']$ is a *local* static subclass of A and hence, is the static A-copy of some $x_0 \in A$. But since $A \models x_0 \subseteq U_A$ it follows that $x_0 \in U_A$ and hence, $A' \models F(x_0) = x'$, which contradicts the fact that $A' \models x' \notin F[U_A]$.

Thus F is onto, is an isomorphism, and by its uniqueness we get the result claimed. □

As an application of this result, assume $A \subseteq A'$ is an extension of universes and that $\kappa \in A$ satisfies

$$A \models \kappa \text{ is a cardinal.}$$

Then at the very least we have

$$A \models \kappa \in U_A$$

and thus, to κ canonically corresponds a $\kappa' \in U_{A'}$ for which, obviously,

$$A' \models \kappa' \text{ is a cardinal.}$$

In this manner, we see that to any of A's cardinal numbers κ there canonically and bi-uniquely corresponds a cardinal number κ' of A'. In effect, the cardinal numbers κ and κ' will represent the "same" level of cardinality within their respective frames of reference.

This suggests the following definition:

DEFINITION 12.1. *Let A be a universe with cardinal number κ. We call an extension $A \subseteq A'$ of universes κ-saturated if to κ corresponds the cardinal number κ' in A' and it is the case that*

$$A' \models (\forall w \subseteq \overline{A})[card(w) < \kappa' \ \& \ w \ satisfies \ f.i.p. \ \to \cap w \neq \emptyset],$$

where the clause "w satisfies f.i.p." abbreviates

$$(\forall w_0)[w_0 \subseteq w \ \& \ w_0 \ finite \ \to \cap w_0 \neq \emptyset].$$

This definition implies a saturation property for extensions of *EST*-universes which is similar to the saturation axioms in the "non"standard set theories we studied previously. Indeed, we may imagine the inclusion of static *EST*-classes $A \subseteq \overline{A} \subseteq A'$ to be playing the role of the previous inclusions $S \subseteq I \subseteq E \ (= V)$ of standard, internal and external universes. If one takes a formula $\varphi(x, y, \mathbf{z})$ from the language $\mathbf{L}(ZFC)$ whose free variables are among x, y and $\mathbf{z} = z_1, \dots, z_n$, then the previous saturation axioms had the form

$$(\forall^I \mathbf{z})(\forall^{small} D \subseteq I)[(\forall^{Fin} d \subseteq D)(\exists^I y)(\forall^I x \in d) \, {}^I\varphi(x, y, \mathbf{z})$$
$$\rightarrow (\exists^I y)(\forall^I x \in D) \, {}^I\varphi(x, y, \mathbf{z})].$$

Assume $A \subseteq \overline{A} \subseteq A'$ is a κ-saturated extension of universes and temporarily call a static subclass $D \subseteq A'$ "small" if (with the obvious meaning)

$$A' \models card(D) < \kappa',$$

where κ' is the A'-cardinal number corresponding to κ. The corresponding saturation property, in this context, would of course be

$$A' \models (\forall^{\overline{A}} \mathbf{z})(\forall^{small} D \subseteq \overline{A})[(\forall^{Fin} d \subseteq D)(\exists^{\overline{A}} y)(\forall^{\overline{A}} x \in d) \, {}^{\overline{A}}\varphi(x, y, \mathbf{z})$$
$$\rightarrow (\exists^{\overline{A}} y)(\forall^{\overline{A}} x \in D) \, {}^{\overline{A}}\varphi(x, y, \mathbf{z})].$$

Our next proposition shows this to be the case. It will be helpful, however, to first restate matters slightly. If we let $B' \subseteq A'$ satisfy

$$A' \models B' = \{< x, \mathbf{z}, y >\in \overline{A} : \, {}^{\overline{A}}\varphi(x, y, \mathbf{z})\},$$

then by the transfer principle for *EST*-universes, if $B = B' \cap A$, then

$$A' \models B = \{< x, \mathbf{z}, y >\in A : \, {}^A\varphi(x, y, \mathbf{z})\}$$

and $\overline{B} = B'$. Assume $D \subseteq \overline{A}$ is "small". If we fix a choice of $\mathbf{z} \in \overline{A}$ and let $D' \subseteq A'$ be the static class

$$\{< x, \mathbf{z} >: x \in D\},$$

then D' is also "small". For each choice of set $x' \, =< x, \mathbf{z} >\in D'$, let us abbreviate by $\overline{B}_{x'}$ the static class

$$\{y :< x, \mathbf{z}, y >\in \overline{B}\} \subseteq A'.$$

To say that

$$A' \models (\forall^{Fin} d \subseteq D)(\exists^{\overline{A}} y)(\forall^{\overline{A}} x \in d) \, {}^{\overline{A}}\varphi(x, y, \mathbf{z})$$

amounts to saying (with the obvious meaning) that

$$A' \models \{(\overline{B})_{x'}\}_{x' \in D'} \text{ satisfies f.i.p.,}$$

and to say that

$$A' \models (\exists^{\overline{A}} y)(\forall^{\overline{A}} x \in D) \, {}^{\overline{A}}\varphi(x, y, \mathbf{z}),$$

is the same as saying that

$$A' \models \bigcap \{(\overline{B})_{x'}\}_{x' \in D'} \neq \emptyset.$$

Also, the (local) static class $D' \subseteq A'$ — as a "collection of indices" — can just as well be replaced by the A'-set w whose static A'-copy is D'.

This motivates the following proposition (interpreted in the obvious manner) which shows that Definition 12.1 implies for extensions of EST-universes the sort of saturation property we have seen in previous "non"standard set theories.

PROPOSITION 12.3. *Suppose $A \subseteq A'$ is a κ-saturated extension of universes and let the cardinal κ' in A' correspond to the cardinal κ in A. Then*

$$A' \models (\forall B \subseteq A)(\forall w \subseteq \overline{A})[card(w) < \kappa' \,\&\, \{(\overline{B})_x\}_{x \in w} \text{ satisfies f.i.p.}$$
$$\rightarrow \bigcap \{(\overline{B})_x\}_{x \in w} \neq \emptyset].$$

PROOF. Assume the hypotheses concerning A, A', κ and κ'. Let $B \subseteq A$ be a static class and $p \in A'$ an element for which

$$A' \models p \subseteq \overline{A} \,\&\, card(p) < \kappa' \,\&\, \{(\overline{B})_x\}_{x \in p} \text{ satisfies f.i.p.}$$

We need to argue that

$$A' \models \bigcap \{(\overline{B})_x\}_{x \in p} \neq \emptyset.$$

Retaining the notation developed in the proof of Proposition 12.1, A'-collection will give us some $\alpha \in On_{A'}$ for which

$$A' \models \{(W_\alpha \cap \overline{B})_x\}_{x \in p} \text{ satisfies f.i.p.}$$

Now A'-collection also gives us an $r \in A'$ for which (obvious meaning)

$$A' \models r = \{(\widehat{W_\alpha \cap \overline{B}}^\dagger)_x : x \in p\}.$$

But for any $s \in p \cap A'$, we have

$$(\widehat{W_\alpha \cap \overline{B}}^\dagger)_s = (\widehat{W_\alpha \cap \overline{B}})_s,$$

from which it is clear that

$$A' \models r \subseteq \overline{A} \,\&\, card(r) < \kappa' \,\&\, r \text{ satisfies f.i.p.}$$

and hence, by the κ-saturation of the extension $A \subseteq A'$, we also have that $A' \models \bigcap r \neq \emptyset$. But with this choice of $r \in A'$, we have

$$A' \models \bigcap r \subseteq \bigcap \{(\overline{B})_x\}_{x \in p},$$

and so we also have

$$A' \models \bigcap \{(\overline{B})_x\}_{x \in p} \neq \emptyset.$$

$\square$

We can now see how some sets in the *EST*-cosmos can be fully seen in a universe, and some cannot. Clearly, any universe A seeing a set x will think it sees all of it, namely, $A \models x \subseteq A$. Typically, however, in an extension $A \subseteq A'$ of universes, x will experience "growth", and $A' \models x \subsetneqq A$ will be the case. On the other hand, if A sees x as finite in the strong sense that

$$\text{for sets } x_1, x_2, \dots, x_n \in A, \ A \models x = \{x_1, x_2, \dots, x_n\},$$

then in any extension $A \subseteq A'$, A' will agree with A about this and $A' \models x \subseteq A$ will hold.

In contrast, suppose that A sees a set x, but sees it as infinite. Then letting ω_A be A's first infinite ordinal, it will turn out that if A has an extension $A \subseteq A'$ which is at least ω_A^+-saturated, then even A' sees more of x than A can. For since A sees x as infinite, A will see an x_0 for which

$$A \models x_0 \subseteq x \ \& \ card(x_0) = \omega.$$

Form the static class

$$B = \{< y, z >: y \in x_0 \cap A \ \& \ z \in (x_0 - \{y\}) \cap A\}$$

and let $w \in A'$ have $x_0 \cap A$ as static A'-copy. Then since

$$A' \models card(w) = \omega < \omega^+$$

and (by familiar arguments)

$$A' \models w \subseteq \overline{A} \ \& \ \{(\overline{B})_y\}_{y \in w} \text{ satisfies f.i.p.,}$$

the hypotheses of Proposition 12.3 apply, and we have

$$A' \models \bigcap \{(\overline{B})_y\}_{y \in w} \neq \emptyset.$$

Thus, A' must see an element of x_0 not seen by A. By transfer applied to $A \models x_0 \subseteq x$, this is also an element of x not seen by A.

Between these two cases, however, an uncertainty arises. It is conceivable for some $x \in A$ that even though $A \models$ "x is finite", it may fail that $x \subseteq A$, or even worse, that in a sufficiently large extension $A \subseteq A'$, it may turn out that $A' \models$ "x is infinite". Indeed, this is why, even though universe extensions $A \subseteq A'$ obey a transfer principle, which may also include saturation, I have hesitated so far to actually call such extensions *enlargements*. These considerations motivate the following definition:

DEFINITION 12.2. *And extension $A \subseteq A'$ of universes is an* enlargement *if*

$$A' \models (\forall x \in A)[\text{"}A \models x \text{ is finite"} \ \to \ x \subseteq A].$$

Enlargements, in this sense, have the nice property that a set seen by the first universe is seen as finite by either universe only if it is seen as finite by both. Moreover, when this is the case, they both essentially agree on how big it is. Indeed, the "finite" cardinals they use to measure it are corresponding cardinals in the sense of Proposition 12.2.

I can now offer the complete list of axioms for the theory *EST*:

Axioms of EST

1) There exists a universe A.
2) For any universe A with cardinal κ, there exists a κ-saturated enlargement $A \subseteq A'$.

Let us see how *EST* fits with the other "non"standard set theories we've studied and how it fulfills the shopping list we drew up earlier. For any *EST*-universe A, its regular part $U_A \subseteq A$ is also a universe and this satisfies the axiom of regularity. These universes (call them *regular*) are characterized by the property that $U_A = A$. Clearly, they model ZFC. Proposition 12.2 strongly suggests that such universes are all canonically interchangeable, and the next chapter offers evidence that we can literally treat them as such.

In each of the "non"standard set theories offered by Nelson, Hrbáček, Kawai, and Fletcher a single copy S of the standard universe is offered. In contrast, we see that *EST*, with its regular universes, offers endless copies. In their own fashion, the non-Fletcher theories each offer, next, a single elementary extension I of their standard universe (the extension being referred to as "the" internal universe), which is either globally saturated at a local level (Nelson with ω-saturation), or locally saturated at a global level (Hrbáček), or globally saturated at a global level (Kawai). Fletcher offers multiple locally saturated elementary extensions (his I_κ's) of his single standard universe, each saturated at a local level κ, with all levels κ available. In *EST*, all universes — "standard" or not — are treated equally and each receives a similar spectrum of locally saturated elementary extensions. Specifically, for given universe A with local saturation level κ (expressed as an A-cardinal), if one chooses a κ-saturated *EST*-enlargement $A \subseteq A'$ then, by Proposition 12.1, the transitive closure $\overline{A}$ of A in A' acts as an "internal" universe elementarily extending the "standard" universe A which is — in the sense of Definition 12.1— locally κ-saturated.

Except for Nelson, each author then offers an external universe E extending their internal universe I (in the case of Fletcher, multiple external universes E_κ are offered extending the separate I_κ's) and, as far as possible, these extensions are designed to model a full ZF^-C. Kawai and Fletcher are able to achieve full ZF^-C for their E's while Hrbáček, for reasons discussed earlier, had to settle with somewhat less. Given a κ-saturated enlargement $A \subseteq \overline{A} \subseteq A'$ within *EST*, the universe A' acts as a universe of externals extending the (acting) universe of internals $\overline{A}$. Not only does A' model full ZF^-C, but clearly — if we ignore local static subclasses of A' which are *non*sets — A' even models a correspondingly full BG^-C^+.

In their own fashion, each author also provides "standardization" of externals. For Nelson and Kawai, this amounts to a strong separation axiom for their standard universe (i.e., separation with respect to formulas in the extended language $\mathbf{L}(NZFC)$). For Hrbáček and Fletcher, it further means for each external set x that $x \cap S$ is modeled by a standard set $^\circ x \in S$ (Kawai can't do this because his externals get too large, e.g., S itself is an external set). Of course in EST, each universe A satisfies such strong standardization: namely, if x is an EST-set, then $x \cap A$ is a local static subclass of A which A recognizes as being local and so, by BG^-C^+-separation, A has (sees) a set $^\circ x$ whose extension (A thinks) is the same as $x \cap A$.

In this manner, the previous "non"standard set theories more or less satisfy — and the current theory EST fully satisfies — items (1) and (2) on our shopping list. None of the previous theories satisfy the last item (3), where notions of standard, internal and external are required to be relativized. Clearly, EST does. A practitioner working in a universe A who needs certain sets to undergo sufficiently saturated internal growth can pass to an appropriate enlargement $A \subseteq \overline{A} \subseteq A'$ and work within $\overline{A}$. If external constructions (e.g., power sets, nonfinitary unions, etc.) are needed, one can pass up into A'. If later it turns out that these new sets are also needing internal growth for the argument to progress, the practitioner can move into a further enlargement $A' \subseteq \overline{A'} \subseteq A''$ and now work in $\overline{A'}$. Of course, this cycle may be repeated many times. In any case, with the arrival of the second enlargement a set x' in A' has already been viewed first as external, then as standard, and then as internal. This phenomenon captures the relativity requested in item (3) on the shopping list.

As a final comparison of EST with Fletcher's $SNST$, we can imagine A, $\overline{A}$ and A' as examples of his S, I_κ and E_κ. However, (for appropriate κ') $\overline{A'}$ and A'' in the second enlargement will *not* correspond to his $I_{\kappa'}$ and $E_{\kappa'}$. This is because Fletcher only allows internals to experience further internal growth, whereas EST permits this for any sets, however they might be being viewed at the moment. In fact, Fletcher's $I_{\kappa'}$ will correspond not to $\overline{A'}$ (the transitive closure in A'' of A'), but to the transitive closure in A'' of $A \subsetneq A'$. Furthermore, Fletcher's $E_{\kappa'}$ will correspond to the proper *sub*class in A'' of sets which are regular over this restricted transitive closure.

In the next chapter we confront the issue of whether EST is "safe" for practitioners.

CHAPTER 13

Conservativity of EST

Recall that an EST-universe A is *regular* if $A = U_A$, i.e., if

$$A \models \text{ the axiom of regularity.}$$

Our central task in this chapter is to prove

THEOREM 13.1. *EST is conservative over ZFC in the following sense: If φ is any sentence of ZFC, then the following are equivalent:*
1) $ZFC \vdash \varphi$
2) $EST \vdash (\exists A)[A$ *is a regular universe* $\& A \models \varphi]$
3) $EST \vdash (\forall A)[A$ *is a regular universe* $\to A \models \varphi]$.

PROOF. I will start out by describing some useful concepts. As always, we continue to work within our AST universe. I call any mapping of sets $f : A \to A'$ a *Z-map* if A and A' are both transitive sets[1] and f is a monomorphism

$$f : (A, \in |_A) \to (A', \in |_{A'})$$

of the associated Z-structures. When this monomorphism is *elementary*, we say the Z-map is also. If the morphism is elementary with respect to restricted formulas we shall say the Z-map $f : A \to A'$ is *restricted elementary*. A Z-map $f : A \to A'$ is *enlarging* if there is a transitive set $\overline{A}$ such that $f[A] \subseteq \overline{A} \subseteq A'$ and for which the induced Z-map

$$f : A \to \overline{A} \subseteq A'$$

is elementary. When we refer to such an enlarging f, a specific such $\overline{A}$ shall be assumed. Although we do not actually assume that $\overline{A}$ is the transitive closure of $f[A]$, the inclusion sequence $f[A] \subseteq \overline{A} \subseteq A'$ here is similar to the inclusion sequence of static EST-classes $A \subseteq \overline{A} \subseteq A'$ discussed in Chapter 12, and for this reason we keep the same notation. Whatever choice of $\overline{A}$ is understood, it is clear that an enlarging Z-map $f : A \to A'$ is also restricted elementary.

---

[1] In this chapter I will freely use capital letters for certain AST-sets, the implication being that relative to V_0-sets they may be large. Often they will be AST-isomorphic to proper V_0-classes.

If κ is a cardinal number, then an enlarging Z-map $f : A \to A'$ is κ-*saturated* if whenever $p' \in A'$ satisfies

$$p' \subseteq \overline{A}, \ \mathrm{card}(p') < \kappa \text{ and } p' \text{ has}$$
$$\text{the finite intersection property,}$$

then $\cap p' \neq \emptyset$.

From the *AST* axiom of superuniversality, it is clear that any directed system

$$\{f_{ij} : A_i \to A_j\}_{i \leq j \in \Gamma}$$

of Z-maps has a Z-map direct limit

$$\{f_i : A_i \to A\}_{i \in \Gamma} .$$

If the original system of Z-maps is *definable* in the sense of Chapter 11[2], then obviously, so will be system of Z-maps making up the direct limit. The same holds for a system of Z-maps which are restricted elementary.

We eventually prove the current theorem by constructing a transitive set M so that $\mathbf{M} = (M, \in |_M)$ models *EST* and has the property that each of its *EST*-universes $A \in M$ satisfies

$$(U_A, \in |_{U_A}) \approx (U_0, \in |_{U_0}).$$

The next concept is crucial to the construction of such a model.

——EST SYSTEMS——

I mean by an *EST-system* any definable directed On_0-indexed system of Z-maps

$$\{f_{\alpha,\alpha'} : \widehat{A}_\alpha \to \widehat{A}_{\alpha'}\}_{\alpha \leq \alpha' \in On_0}$$

such that the following are satisfied:

1) for each $\alpha \in On_0$, $\widehat{A}_\alpha$ is large and for nonlimit α, its definable small subsets are exactly its elements
2) for each $\alpha \in On_0$, the Z-map $f_{\alpha,\alpha+1} : \widehat{A}_\alpha \to \widehat{A}_{\alpha+1}$ is enlarging and $\mathrm{card}(\alpha)^+$-saturated if α is a nonlimit
3) for each limit $\alpha_0 \in On_0$, the system of Z-maps
$$\{f_{\alpha,\alpha'} : \widehat{A}_\alpha \to \widehat{A}_{\alpha'}\}_{\alpha \leq \alpha' < \alpha_0}$$
has $\{f_{\alpha,\alpha_0} : \widehat{A}_\alpha \to \widehat{A}_{\alpha_0}\}_{\alpha < \alpha_0}$ as a direct limit.

I shall abbreviate mention of this third property by saying the On_0-indexed system of Z-maps is *continuous*.

---

[2]Recall that *AST*-set X (or *AST*-system thereof) is *definable* if there is an understood *AST*-isomophism of it onto a V_0-class (or V_0-definable system thereof). As we have seen, the isomorphism allows notions of definability, smallness and largeness to be transported to X. Collection-style arguments on X are then available.

——EST-SYSTEM $\Rightarrow$ EST MODEL——

Given any such *EST*-system, we may construct a model of *EST* in the following manner: We choose a (necessarily definable) direct limit

$$\{f_\alpha : \widehat{A}_\alpha \to M_0\}_{\alpha \in On_0}$$

of the system and abbreviate each $f_\alpha[\widehat{A}_\alpha] \subseteq M_0$ as A_α. We also put

$$M = M_0 \cup \{B \subseteq M_0 : B \text{ is definable and for some } \alpha \in On_0, B \subseteq A_\alpha\}.$$

It is $\mathbf{M} = (M, \in |_M)$ that we claim as our desired model of *EST*. Clearly, the set M is transitive and for each α, A_α reflects restricted formulas for M. Since the latter is transitive, A_α in fact reflects restricted formulas absolutely (i.e., for the entire *AST*-universe).

I will first show that every A_α, for nonlimit α, is an *EST*-universe for $\mathbf{M}$. This will imply that if $\mathbf{M}$ *does* model *EST*, then all its universes will be directed by inclusion. Thus, by Proposition 12.2, all will have isomorphic $(U_A, \in |_{U_A})$'s, so that if any one of these is isomorphic to $(U_0, \in |_{U_0})$, they all are. This will give the equivalence of (2) and (3) in the theorem's statement, at least for the current (presumed) model. As discussed in Chapter 11, if the isomorphisms to $(U_0, \in |_{U_0})$ exist then the equivalence of (1), (2) and (3) are assured in general.

We begin by arguing for each nonlimit α that $(A_\alpha, \in |_{A_\alpha})$ models ZF^-C. For this it suffices to show that its $(\widehat{A}_\alpha, \in |_{\widehat{A}_\alpha})$ does. Since each $p \in \widehat{A}_\alpha$ is a small definable subset of $\widehat{A}_\alpha$, so is each of its subsets $q \subseteq p$. Thus, for each $p \in \widehat{A}_\alpha$, its power set $P(p)$ is a subset of $\widehat{A}_\alpha$ which is definable, clearly is small, and hence, is an element of $\widehat{A}_\alpha$. In other words, $(\widehat{A}_\alpha, \in |_{\widehat{A}_\alpha})$ satisfies the power set axiom and its power sets are true *AST*-power sets. Since $\widehat{A}_\alpha$ is large and contains as elements all its small definable subsets, the remaining ZF^-C axioms for $(\widehat{A}_\alpha, \in |_{\widehat{A}_\alpha})$ are now easily verified.

Next we argue, for nonlimit α, that in every case

$$M_0 \cap P(A_\alpha) = \{p \subseteq A_\alpha : p \text{ finite}\}.$$

Indeed, if $p \subseteq A_\alpha$ is finite, then by ZF^-C axioms and (absolute) restricted reflection, we have $p \in A_\alpha \subseteq M_0$, so the right to left inclusion is immediate. Suppose, conversely, that $p \in M_0$ satisfies $p \subseteq A_\alpha$ although p is not finite. Without loss of generality, we can also assume that both $p \in A_\alpha$ and $\text{card}(p) \leq \text{card}(\alpha)$ are the case. Let $\hat{p} \in \widehat{A}_\alpha$ satisfy $f_\alpha(\hat{p}) = p$ and put

$$\hat{p}' = \{f_{\alpha,\alpha+1}[\hat{p} - \{q\}] : q \in \hat{p}\}.$$

Since $f_\alpha[\hat{p}] = p$ is forced, we have $f_{\alpha,\alpha+1}[\hat{p}] = f_{\alpha,\alpha+1}(\hat{p})$ and $f_{\alpha,\alpha+1}[\hat{p} - \{q\}] = f_{\alpha,\alpha+1}(\hat{p} - \{q\})$, for every $q \in \hat{p}$. Thus the following all hold:

$$\hat{p}' \in \widehat{A}_{\alpha+1}, \quad \text{card}(\hat{p}') \leq \text{card}(\alpha), \quad \cap \hat{p}' = \emptyset,$$
$$\hat{p}' \subseteq \text{the transitive closure of } f_{\alpha,\alpha+1}[\widehat{A}_\alpha] \subseteq \widehat{A}_{\alpha+1}$$
$$\text{and } \hat{p}' \text{ satisfies the finite intersection property.}$$

But this contradicts the fact that the enlarging Z-map $f_{\alpha,\alpha+1} : \widehat{A}_\alpha \to \widehat{A}_{\alpha+1}$ is $\operatorname{card}(\alpha)^+$-saturated, and thus, the left to right inclusion is proved.

For each *non*limit $\alpha \in On_0$, the fact now that all of

$$(A_\alpha, \in |_{A_\alpha}) \text{ models } ZF^-C,$$
$$A_\alpha \text{ is definable and large}$$
$$M_0 \cap P(A_\alpha) = \{p \subseteq A_\alpha : p \text{ finite}\}$$

hold make it routine to show[3] that $A_\alpha \models BG^-C^+_{weak\,ext}$ is the case for $\mathbf{M} = (M, \in |_M)$. The remaining features for A_α to be an EST-universe are also routinely argued.

I now want to show that if $A \in M$ is any of $\mathbf{M}$'s EST-universes, then its $(U_A, \in |_{U_A})$ is isomorphic to $(U_0, \in |_{U_0})$. It suffices to consider the case when $A \in M$ has its $(A, \in |_A)$ isomorphic to some definable large Z-structure $(\widehat{A}, \in |_{\widehat{A}})$ where $\widehat{A}$ is transitive, $(\widehat{A}, \in |_{\widehat{A}})$ models ZF^-C, and $\widehat{A}$'s definable small subsets are exactly its elements. We need only show under these circumstances that $(U_{\widehat{A}}, \in |_{U_{\widehat{A}}})$ must be isomorphic to $(U_0, \in |_{U_0})$. But by definability and largeness, this $(U_{\widehat{A}}, \in |_{U_{\widehat{A}}})$ will be isomorphic to an extensional well-founded Z-structure (W, E) where W and E are proper V_0-classes. Since W and E are large and the subsets of W which are V_0-sets are exactly those E-represented by elements of W, it follows by the Mostowski Collapsing Lemma (*global* form) that (W, E) is isomorphic to $(U_0, \in |_{U_0})$. Thus, so is $(U_{\widehat{A}}, \in |_{U_{\widehat{A}}})$.

From this it follows that each extension $A \subseteq A'$ of $\mathbf{M}$'s EST-universes is already an enlargement. To see this, it will suffice to show for any of $\mathbf{M}$'s EST-universes A that

$$A \models p \text{ finite} \Rightarrow p \text{ finite}.$$

Now if

$$A \models p \text{ finite},$$

then there exists $f, n \in A$ such that $n \in \omega_A$ $(= A$'s version of $\omega)$ and

$$A \models f : p \to n \text{ bijectively},$$

which, by (absolute) restricted reflection, implies that f is a one to one map $p \to n$ and thus, that $p \cap A$ is bijective with $n \cap A = n \cap U_A$. But the latter (via the isomorphism of U_A with U_0) is bijective with a finite ordinal and so $p \cap A$ itself is finite. Thus $p_1, p_2, \ldots, p_n \in A$ exist for which

$$A \models p = \{p_1, p_2, \ldots, p_n\},$$

and by (absolute) restricted reflection, this shows p is indeed finite.

We are now poised to show that $\mathbf{M}$ is an actual model of EST. Let $A \in M$ be any EST-universe for $\mathbf{M}$ and let $\kappa \in A$ be one of its infinite cardinals. We already know that $(U_A, \in |_{U_A})$ is canonically isomorphic to $(U_0, \in |_{U_0})$. Let $\kappa_0 \in U_0$ be the cardinal corresponding to κ. Pick nonlimit $\alpha \in On_0$ large enough so that

---

[3]Use the fact that M-classes are definable.

$A \subseteq A_\alpha$ and $\kappa_0 \leq \mathrm{card}(\alpha)^+$. We want to show that for **M**, the enlargement $A \subseteq A_{\alpha+1}$ is κ-saturated, in the sense of *EST*-theory. Let $\kappa' \in A_{\alpha+1}$ be the $A_{\alpha+1}$-cardinal number corresponding to $\kappa \in A$ (and hence to $\kappa_0 \in On_0$). Choose $p \in A_{\alpha+1}$ such that each of

$$A_{\alpha+1} \models \mathrm{card}(p) < \kappa'$$
$$A_{\alpha+1} \models p \subseteq \text{ the transitive closure of } A$$
$$A_{\alpha+1} \models p \text{ satisfies f.i.p.}$$

hold. Letting $\alpha' \in On_{A_{\alpha+1}}$ correspond to $\alpha \in On_0$, we have for some $f' \in A_{\alpha+1}$ that

$$A_{\alpha+1} \models f' : p \to \alpha' \text{ injectively,}$$

so by (absolute) restricted reflection, $f' : p \to \alpha'$ is a one to one map and hence,

$$\mathrm{card}(p \cap A_{\alpha+1}) \leq \mathrm{card}(\alpha' \cap A_{\alpha+1}) = \mathrm{card}(\alpha \cap On_0) = \mathrm{card}(\alpha).$$

Similarly, since $A_{\alpha+1}$-finiteness is actual finiteness the collection

$$\{q \cap A_{\alpha+1} : q \in p \cap A_{\alpha+1}\}$$

itself satisfies f.i.p. Now choose transitive $f_{\alpha,\alpha+1}[\widehat{A}_\alpha] \subseteq \overline{\widehat{A}}_\alpha \subseteq \widehat{A}_{\alpha+1}$ with respect to which $f_{\alpha,\alpha+1} : \widehat{A}_\alpha \to \widehat{A}_{\alpha+1}$ is understood to be enlarging and $\mathrm{card}(\alpha)^+$-saturated and also choose $\hat{p} \in A_{\alpha+1}$ so that $f_{\alpha+1}(\hat{p}) = p$. Then we have

$$\hat{p} \subseteq \overline{\widehat{A}}_\alpha \text{ (since } A \subseteq A_\alpha\text{)}$$
$$\mathrm{card}(\hat{p}) < \mathrm{card}(\alpha)^+$$
$$\hat{p} \text{ satisfies f.i.p.,}$$

and by the $\mathrm{card}(\alpha)^+$-saturation of $f_{\alpha,\alpha+1}$, it follows that $\cap \hat{p} \neq \emptyset$. Clearly, this implies

$$A_{\alpha+1} \models \cap p \neq \emptyset$$

and therefore, that the enlargement $A \subseteq A_{\alpha+1}$ is κ-saturated in the sense of *EST*-theory. More generally, we see that **M** satisfies all *EST* axioms.

——PARTIAL EST-SYSTEMS, COHERENT FAMILIES——

To finish the proof of our theorem, we have to show the existence of some *EST*-system. Crucial to this project will be the following two (*decidedly* more technical) concepts.

For any ordinal $\alpha \in On_0$, we will call a *partial EST*-system of *level* α any pair of V_0-sets

$$\{f^\beta_{\alpha',\alpha''} : A^\beta_{\alpha'} \to A^\beta_{\alpha''}\}_{\alpha' \leq \alpha'' < \alpha, \beta < \alpha}$$
$$\text{and } \{\tau^\beta_{\alpha'}\}_{\alpha'+1, \beta < \alpha}$$

where the first is a system of Z-maps and the second a family of internal domain structures such that

1) for all $\alpha' < \alpha$, the collection of sets $\{A_{\alpha'}^{\beta}\}_{\beta<\alpha}$ is increasing and continuous and if $\beta + 1 < \alpha$ then for any *non*limit $\alpha' < \alpha$, $P(A_{\alpha'}^{\beta}) \subseteq A_{\alpha'}^{\beta+1}$ holds

2) for all $\alpha' \leq \alpha'' < \alpha$ and $\beta < \beta' < \alpha$, the map $f_{\alpha',\alpha''}^{\beta'}$ extends the map $f_{\alpha',\alpha''}^{\beta}$

3) for each $\beta < \alpha$, the system of Z-maps $\{f_{\alpha',\alpha''}^{\beta} : A_{\alpha'}^{\beta} \to A_{\alpha''}^{\beta}\}_{\alpha' \leq \alpha'' < \alpha}$ is directed and, in the sense used when defining full *EST*-systems, is *continuous*

4) For $\alpha' + 1, \beta < \alpha$, $\tau_{\alpha'}^{\beta}$ is an internal domain structure on some transitive set $\overline{A}_{\alpha'}^{\beta} \subseteq A_{\alpha'+1}^{\beta}$ so that

   a) the core $= {}^{\circ}\overline{A}_{\alpha'}^{\beta} = f_{\alpha',\alpha'+1}^{\beta}[A_{\alpha'}^{\beta}]$

   b) $\in |_{\overline{A}_{\alpha'}^{\beta}}$ is $\tau_{\alpha'}^{\beta}$-clopen

   c) the sequence $\{\overline{A}_{\alpha'}^{\beta'}\}_{\beta' \leq \beta}$ is increasing and continuous, consists of $\tau_{\alpha'}^{\beta}$-open subsets of $\overline{A}_{\alpha'}^{\beta}$ where, for $\beta' < \beta$, the $\tau_{\alpha'}^{\beta}$-internal subdomain structure on $\overline{A}_{\alpha'}^{\beta'}$ is that given by $\tau_{\alpha'}^{\beta'}$, and for *non*limit $\beta' < \beta$, $\overline{A}_{\alpha'}^{\beta'}$ is $\tau_{\alpha'}^{\beta}$-clopen and $\tau_{\alpha'}^{\beta}$-local

   d) for each $p \in \overline{A}_{\alpha'}^{\beta}$, $\mathrm{rank}(p) \leq \mathrm{card}(\alpha')^4$

   e) for each $\beta' < \beta$ and each clopen $Z \subseteq \overline{A}_{\alpha'}^{\beta'+1}$, if α' is a nonlimit and $\mathrm{card}({}^{\circ}Z) \leq \mathrm{card}(\alpha')$, then Z is compact and $\mathrm{card}(\alpha')^+$-saturated.

I also call a *coherent On_0-indexed family* of partial *EST*-systems any pair of proper V_0-classes

$$\{f_{\alpha',\alpha''}^{\beta} : A_{\alpha'}^{\beta} \to A_{\alpha''}^{\beta}\}_{\alpha' \leq \alpha'' \in On_0, \beta \in On_0}$$
$$\text{and } \{\tau_{\alpha'}^{\beta}\}_{\alpha',\beta \in On_0}$$

such that for each $\alpha \in On_0$ the subcollections

$$\{f_{\alpha',\alpha''}^{\beta} : A_{\alpha'}^{\beta} \to A_{\alpha''}^{\beta}\}_{\alpha' \leq \alpha'' < \alpha, \beta < \alpha}$$
$$\text{and } \{\tau_{\alpha'}^{\beta}\}_{\alpha'+1,\beta < \alpha}$$

form a partial *EST*-system of level α.

——COHERENT FAMILY $\Rightarrow$ EST SYSTEM——

The relevance of the second concept is that from a coherent On_0-indexed family of partial *EST*-systems we can actually construct a full *EST*-system. Indeed, let

$$\{f_{\alpha',\alpha''}^{\beta} : A_{\alpha'}^{\beta} \to A_{\alpha''}^{\beta}\}_{\alpha' \leq \alpha'' \in On_0, \beta \in On_0}$$
$$\text{and } \{\tau_{\alpha'}^{\beta}\}_{\alpha',\beta \in On_0}$$

---

[4]Recall Definition 7.1.

be such a coherent family of partial systems. We put for each $\alpha \in On_0$

$$A_\alpha = \bigcup_{\beta \in On_0} A_\alpha^\beta \text{ and } \overline{A}_\alpha = \bigcup_{\beta \in On_0} \overline{A}_\alpha^\beta \subseteq A_{\alpha+1}$$

and for $\alpha \leq \alpha' \in On_0$, we put

$$f_{\alpha,\alpha'} = \bigcup_{\beta \in On_0} f_{\alpha,\alpha'}^\beta : A_\alpha \to A_{\alpha'}.$$

It is the collection

$$\{f_{\alpha,\alpha'} : A_\alpha \to A_{\alpha'}\}_{\alpha \leq \alpha' \in On_0}$$

that I claim is a full *EST*-system. The construction of the system of Z-maps clearly shows it is definable. In fact, the only requirement for being an *EST*-system that is nonroutine, is to check that each Z-map

$$f_{\alpha,\alpha+1} : A_\alpha \to A_{\alpha+1}$$

is enlarging and, for nonlimit α, is also $\mathrm{card}(\alpha)^+$-saturated.

Clearly, for each $\alpha \in On_0$,

$$\{\overline{A}_\alpha^\beta, \tau_\alpha^\beta\}_{\beta \in On_0}$$

is an increasing continuous sequence of internal domain extensions. For each $\beta' < \beta \in On_0$, $\overline{A}_\alpha^{\beta'}$ is open in $\overline{A}_\alpha^\beta$ and indeed, local and clopen if β' is a nonlimit. This system clearly induces a unique internal domain structure τ_α on $\overline{A}_\alpha$ whose core is

$$^\circ \overline{A}_\alpha = f_{\alpha,\alpha+1}[A_\alpha].$$

By construction $\in|_{\overline{A}_\alpha}$ is τ_α-clopen.

Since A_α and $\overline{A}_\alpha$ are transitive proper V_0-classes an argument using both internal domain transfer and the generalized ZF^-C reflection principle easily shows that the map

$$f_{\alpha,\alpha+1} : A_\alpha \to A_{\alpha+1}$$

is enlarging with respect to $\overline{A}_\alpha$.

We need to also show it is $\mathrm{card}(\alpha)^+$-saturated, when α is a nonlimit. Suppose $p \in A_{\alpha+1}$ satisfies

$$p \subseteq \overline{A}_\alpha, \quad \mathrm{card}(p) \leq \mathrm{card}(\alpha) \text{ and}$$
$$p \text{ has the finite intersection property.}$$

We need to argue that $\cap p \neq \emptyset$. Obviously, we can assume that α is infinite. Since $p \in A_{\alpha+1}$ we have that p is a V_0-set and hence, by V_0-collection $p \subseteq \overline{A}_\alpha^{\beta+1}$ for some $\beta \in On_0$. By construction, each $q \in p$ has $\mathrm{rank}(q) \leq \mathrm{card}(\alpha)$ and thus lies in a clopen $Z_q \subseteq \overline{A}_\alpha^{\beta+1}$ for which $\mathrm{card}(^\circ Z_q) \leq \mathrm{card}(\alpha)$. Letting $^\circ Z = \bigcup_{q \in p} {}^\circ Z_q$ have closure Z we get that $p \subseteq Z \subseteq \overline{A}_\alpha^{\beta+1}$ where Z is clopen and $\mathrm{card}(^\circ Z) \leq \mathrm{card}(\alpha)$. But then Z, by construction, is compact and $\mathrm{card}(\alpha)^+$-saturated, and

since we can choose Z so that $\{q \cap Z : q \in p\}$ continues to have f.i.p., we see that
that $\cap p \neq \emptyset$ is forced.

Thus each $f_{\alpha,\alpha+1} : A_\alpha \to A_{\alpha+1}$ is enlarging, $\operatorname{card}(\alpha)^+$-saturated when α is
a nonlimit, and we have achieved a full *EST*-system from the given coherent
family of partial *EST*-systems.

——PARTIAL EST-SYSTEMS $\Rightarrow$ COHERENT FAMILY——

Obviously, to finish the proof of our theorem, I need to show that coherent
families of partial *EST*-systems exist. Since V_0-global choice is available, it
suffices for this to show that any partial *EST*-system of level $\alpha \in On_0$ extends
to some partial *EST*-system of level $\alpha + 1$,

This is our final task. Assume we are given V_0-sets

$$\{f_{\alpha',\alpha''}^\beta : A_{\alpha'}^\beta \to A_{\alpha''}^\beta\}_{\alpha' \le \alpha'' < \alpha, \beta < \alpha}$$
$$\text{and } \{\tau_{\alpha'}^\beta\}_{\alpha'+1, \beta < \alpha}$$

which make up a partial *EST*-system of level $\alpha \in On_0$. If α is a limit ordinal,
we can set for $\alpha' \le \alpha'' < \alpha$,

$$A_{\alpha'}^\alpha = \bigcup_{\beta < \alpha} A_{\alpha'}^\beta$$
$$f_{\alpha',\alpha''}^\alpha = \bigcup_{\beta < \alpha} f_{\alpha',\alpha''}^\beta : A_{\alpha'}^\alpha \to A_{\alpha''}^\alpha$$

and choose a V_0-set Z-map direct limit

$$\{f_{\alpha',\alpha}^\alpha : A_{\alpha'}^\alpha \to A_\alpha^\alpha\}_{\alpha' < \alpha}$$

of the directed system

$$\{f_{\alpha',\alpha''}^\alpha : A_{\alpha'}^\alpha \to A_{\alpha''}^\alpha\}_{\alpha' \le \alpha'' < \alpha} \, .$$

For every $\beta < \alpha$, let

$$f_{\alpha',\alpha}^\beta = \text{ the restriction of } f_{\alpha',\alpha}^\alpha \text{ to } A_{\alpha'}^\beta$$
$$A_\alpha^\beta = \bigcup_{\alpha' < \alpha} f_{\alpha',\alpha}^\beta [A_{\alpha'}^\beta] \, .$$

The V_0-set

$$\{f_{\alpha',\alpha''}^\beta : A_{\alpha'}^\beta \to A_{\alpha''}^\beta\}_{\alpha' \le \alpha'' \le \alpha, \beta \le \alpha}$$

will make up the Z-map part of our extending partial *EST*-system. For the
internal domains, since $\alpha' + 1$, $\beta < \alpha + 1$ imply that $\alpha' + 1 < \alpha$ [5] and $\beta \le \alpha$, we
need only specify appropriate internal domain structures $\tau_{\alpha'}^\alpha$ for $\alpha' < \alpha$. But by
assumptions, the family

$$\{\overline{A}_{\alpha'}^\beta, \ \tau_{\alpha'}^\beta\}_{\beta < \alpha}$$

is an increasing continuous sequence of internal domain extensions and, as before,
induces a unique internal domain structure $\tau_{\alpha'}^\alpha$, now on the V_0-set limit $\overline{A}_{\alpha'}^\alpha$.
With these choices a V_0-set of internal domain structures

$$\{\tau_{\alpha'}^\beta\}_{\alpha'+1, \beta \le \alpha}$$

---

[5] Strict inequality since α is a limit.

is now specified for the other component of our extending partial *EST*-system. Checking that these two V_0-sets do make a partial *EST*-system of level $\alpha + 1$, extending the original, is routine.

The less routine case is when the level of the original partial *EST*-system is a sucessor $\alpha = \alpha_0 + 1$. Visualizing this system as an α-by-α square array, we shall fill it out to a new system of order $\alpha + 1$ by inductively adding components along the top moving from left to right and then, more or less simultaneously, add the final right hand column in one construction.

Let us call a *ridge line extension* of *length* $\hat{\alpha} \leq \alpha$ any pair of V_0-sets

$$\{f^\alpha_{\alpha',\alpha''} : A^\alpha_{\alpha'} \to A^\alpha_{\alpha''}\}_{\alpha' \leq \alpha'' < \hat{\alpha}}$$
$$\text{and } \{\tau^\alpha_{\alpha'}\}_{\alpha'+1 < \hat{\alpha}}$$

where the first is a continuous directed system of Z-maps and the second a family of internal domain structures such that

1) for all $\alpha' < \hat{\alpha}$, $A^{\alpha_0}_{\alpha'} \subseteq A^\alpha_{\alpha'}$ and if α' is a nonlimit, then $P(A^{\alpha_0}_{\alpha'}) \subseteq A^\alpha_{\alpha'}$

2) for all $\alpha' \leq \alpha'' < \hat{\alpha}$ the map $f^\alpha_{\alpha',\alpha''}$ extends the map $f^{\alpha_0}_{\alpha',\alpha''}$

3) For $\alpha' + 1 < \hat{\alpha}$, $\tau^\alpha_{\alpha'}$ is an internal domain structure on some transitive set $\overline{A}^\alpha_{\alpha'} \subseteq A^\alpha_{\alpha'+1}$ so that

   a) the core $= {}^\circ \overline{A}^\alpha_{\alpha'} = f^\alpha_{\alpha',\alpha'+1}[A^\alpha_{\alpha'}]$

   b) $\in |_{\overline{A}^\alpha_{\alpha'}}$ is $\tau^\alpha_{\alpha'}$-clopen

   c) $\overline{A}^{\alpha_0}_{\alpha'} \subseteq \overline{A}^\alpha_{\alpha'}$ is $\tau^\alpha_{\alpha'}$-open with its $\tau^\alpha_{\alpha'}$-internal subdomain structure given by $\tau^{\alpha_0}_{\alpha'}$ and, if α_0 is a nonlimit, then $\overline{A}^{\alpha_0}_{\alpha'}$ is also $\tau^\alpha_{\alpha'}$-clopen and $\tau^\alpha_{\alpha'}$-local

   d) for each $p \in \overline{A}^\alpha_{\alpha'}$, $\text{rank}(p) \leq \text{card}(\alpha')$

   e) if α' is a nonlimit, then for each clopen $Z \subseteq \overline{A}^\alpha_{\alpha'}$ if $\text{card}(^\circ Z) \leq \text{card}(\alpha')$, then Z is compact and $\text{card}(\alpha')^+$-saturated.

To get a ridge line extension of length 1, choose any transitive V_0-set A^α_0 so that

$$A^{\alpha_0}_0 \subseteq P(A^{\alpha_0}_0) \subseteq A^\alpha_0,$$

and let

$$f^\alpha_{0,0} : A^\alpha_0 \to A^\alpha_0$$

be the identity map. No τ's are needed in this case,

Assume by induction that a ridge line extension of length $\hat{\alpha} < \alpha$ has been constructed. I describe now how to add to it to get a ridge line extension of length $\hat{\alpha} + 1 \leq \alpha$,

If $\hat{\alpha}$ is a limit ordinal, use the axiom of superuniversality for the *AST*-universe (and hence for the elementary subuniverse V_0) to find a V_0-set Z-map direct limit

$$\{f^\alpha_{\alpha',\hat{\alpha}} : A^\alpha_{\alpha'} \to A^\alpha_{\hat{\alpha}}\}_{\alpha' < \hat{\alpha}}$$

of the system

$$\{f^\alpha_{\alpha',\alpha''} : A^\alpha_{\alpha'} \to A^\alpha_{\alpha''}\}_{\alpha' \leq \alpha'' < \hat{\alpha}}$$

which *also* satisfies (*full* use of superuniversality now)

$$A_{\hat{\alpha}}^{\alpha_0} \subseteq A_{\hat{\alpha}}^{\alpha} \text{ and for } \alpha' < \hat{\alpha},$$
$$f_{\alpha',\hat{\alpha}}^{\alpha} : A_{\alpha'}^{\alpha} \to A_{\hat{\alpha}}^{\alpha} \text{ extends } f_{\alpha',\hat{\alpha}}^{\alpha_0} : A_{\alpha'}^{\alpha_0} \to A_{\hat{\alpha}}^{\alpha_0}.$$

Since $\hat{\alpha}$ is a limit, no new τ's need to be chosen and we have a ridge line extension of length $\hat{\alpha} + 1$.

If $\hat{\alpha} = \hat{\alpha}_0 + 1$ is a successor, we proceed as follows. Choose any internal domain $X \in V_0$ with core $^\circ X = A_{\hat{\alpha}_0}^{\alpha_0}$ such that if we let

$$E = \text{ the } X\text{-closure of } \in |_{A_{\hat{\alpha}_0}^{\alpha_0}},$$

then there is an isomorphism of Z-structures

$$f : (X, E) \approx (\overline{A}_{\hat{\alpha}_0}^{\alpha_0}, \in |_{\overline{A}_{\hat{\alpha}_0}^{\alpha_0}})$$

which extends the map

$$f_{\hat{\alpha}_0,\hat{\alpha}_0+1}^{\alpha_0} : A_{\hat{\alpha}_0}^{\alpha_0} \to \overline{A}_{\hat{\alpha}_0}^{\alpha_0} \subseteq A_{\hat{\alpha}_0+1}^{\alpha_0},$$

and which is also an isomorphism of internal domains. We also require of X that

$$X \cap A_{\hat{\alpha}_0}^{\alpha} = A_{\hat{\alpha}_0}^{\alpha_0}.$$

Use Proposition 7.2 to extend X to a *local* internal domain $X' \in V_0$ whose core $^\circ X'$ is $A_{\hat{\alpha}_0}^{\alpha}$, and in which $X \subseteq X'$ is open. The method of construction of X' which was used in the proof of Proposition 7.2 guarantees that the point rank, local compactness, and saturation assumptions made for $\tau_{\hat{\alpha}_0}^{\alpha_0}$ continue to hold for the internal domain X'. Let

$$E' = \text{ the } X'\text{-closure of } \in |_{A_{\hat{\alpha}_0}^{\alpha}}.$$

Since X' is local and $(^\circ X', \in |_{\circ X'})$ is extensional, the transfer principle for internal domains shows that (X', E') is also extensional. Appealing again to (*full*) V_0-superuniversality, we can extend the map f to an isomorphism of Z-structures

$$f' : (X', E') \approx (\overline{A}_{\hat{\alpha}_0}^{\alpha}, \in |_{\overline{A}_{\hat{\alpha}_0}^{\alpha}})$$

where $\overline{A}_{\hat{\alpha}_0}^{\alpha}$ is some transitive V_0-set extending $\overline{A}_{\hat{\alpha}_0}^{\alpha_0}$. Let $\tau_{\hat{\alpha}_0}^{\alpha}$ be the internal domain structure on $\overline{A}_{\hat{\alpha}_0}^{\alpha}$ so that f' is also an isomorphism of internal domains. Next, let $A_{\hat{\alpha}}^{\alpha}$ be any choice of a transitive V_0-set such that

$$\overline{A}_{\hat{\alpha}_0}^{\alpha} \cup P(A_{\hat{\alpha}}^{\alpha_0}) \subseteq A_{\hat{\alpha}}^{\alpha},$$

and for any $\alpha' < \hat{\alpha}_0$, let

$$f_{\alpha',\hat{\alpha}}^{\alpha} = \text{ the composition of } f' \text{ with } f_{\alpha',\hat{\alpha}_0}^{\alpha}.$$

These choices added to the assumed ridge line extension of length $\hat{\alpha}$ clearly give a new ridge line extension of length $\hat{\alpha} + 1$.

By induction, we arrive at a ridge line extension

$$\{f^\alpha_{\alpha',\alpha''} : A^\alpha_{\alpha'} \to A^\alpha_{\alpha''}\}_{\alpha' \le \alpha'' < \alpha}$$
$$\text{and } \{\tau^\alpha_{\alpha'}\}_{\alpha'+1 < \alpha}$$

of length α. But for one right hand entry this will amount to the top row of the new $(\alpha + 1)$-by-$(\alpha + 1)$ array which will be the partial *EST*-system of order $\alpha + 1$ extending the original of order α. To finish matters, we need to construct the new array's entire right hand column.

Using Theorem 7.5 and the fact that any set with the discrete topology is a local internal domain, we are able to pick some $\mathrm{card}(\alpha_0)^+$-saturated local internal domain $X' \in V_0$ with core ${}^\circ X' = A^\alpha_{\alpha_0}$. If we let

$$X = \{p \in X' : rank(p) \le card(\alpha_0)\} \subseteq X',$$

then clearly, this an open subinternal domain of X' with the same core, which is also local. If we let

$$E = \text{ the } X\text{-closure of } \in |_{A^\alpha_{\alpha_0}},$$

then by transfer, the Z-structure (X, E) is extensional. Pick set realization

$$f : (X, E) \approx (\overline{A}^\alpha_{\alpha_0}, \in |_{\overline{A}^\alpha_{\alpha_0}})$$

where $\overline{A}^\alpha_{\alpha_0} \in V_0$ is transitive. Give $\overline{A}^\alpha_{\alpha_0}$ the internal domain structure $\tau^\alpha_{\alpha_0}$ which makes f an isomorphism of internal domains. For $\beta \le \alpha_0$, let $\overline{A}^\beta_{\alpha_0} = f[X^\beta]$ where if β is a nonlimit

$$X^\beta = \text{ the } X\text{-closure of } A^\beta_{\alpha_0},$$

and if β is a limit

$$X^\beta = \bigcup_{\beta' < \beta} X^{\beta'+1}.$$

Let $\tau^\beta_{\alpha_0}$ be the $\tau^\alpha_{\alpha_0}$-subinternal domain structure on its open subset $\overline{A}^\beta_{\alpha_0} \subseteq \overline{A}^\alpha_{\alpha_0}$.

Finally, use global V_0-choice to construct an increasing continuous sequence $\{A^\beta_\alpha\}_{\beta \le \alpha}$ of transitive V_0-sets such that for each $\beta \le \alpha_0$

$$\overline{A}^\beta_{\alpha_0} \cup P(A^\beta_\alpha) \subseteq A^{\beta+1}_\alpha.$$

For α', $\beta \le \alpha_0$, put

$$f^\beta_{\alpha',\alpha} = \text{ the composition of } f \text{ with } f^\beta_{\alpha',\alpha_0}.$$

It is now routine to check that the foregoing choices create a partial *EST*-system of level $\alpha + 1$ which extends the original partial system of level α.

$\square$

Discussion. For reasons of aesthetics, I presented EST with only two axioms. In fact, there are many further axioms that can be conservatively added to EST in the sense that Theorem 13.1 remains valid. The proof I have just given of this theorem shows it may be further assumed as axioms for EST that the totality of all EST classes is extensional, that static classes are closed under finite unions, that any static class may be extended to a full EST-universe and so on. The proof also allows extra freedom at various stages during which the EST model is being constructed. Specifically, this occurs whenever the choice of a transitive V_0-set $A_{\alpha'+1}^{\beta+1}$ is made which extends both $\overline{A}_{\alpha'}^{\beta}$ and $P(A_{\alpha'+1}^{\beta})$. Using global V_0-choice, these choices can be orchestrated, for example, to insure that the EST-universe $A_{\alpha'+1}$ satisfies Boffa's axiom of superuniversality and hence, that every EST-universe may have arbitrarily saturated enlargements which also satisfy this axiom. The interested reader can easily discover further possibilities.

CHAPTER 14

Concluding Remarks

At this point, our prototype EST as an "ultimate" vehicle for "non"standard mathematical practice has been fully described and shown to be "safe" (i.e., conservative over ZFC). It has been assembled and laid out on the table. The greater mathematical community may now rush over and use it.

I speak ironically of course. Of all the various proposed "non"standard set theories, only Nelson's IST found any use amongst practitioners of "non"standard mathematics. Practitioners have tended to stick to Robinson's original enlargements or, if using IST, have proceeded informally adding external sets as the need arose. On the one hand, proposers of new "non"standard set theories have been routinely expected to tailor their vehicle to the "needs" of practitioners, to argue the vehicle's practical advantages for these people, and to show by examples how they can readily use it. On the other hand, a view has become popular that new work in the foundations of "non"standard mathematics is unnecessary, that enough is already known for practitioners to safely pursue the art. At the same time, although it doesn't receive much comment, it can be said that if "non"standard mathematics might have been a revolution among the wider mathematical community, then this revolution has stalled. A subcommunity of practitioners does exist, but its numbers have stabilized and it is not wide spread.

Reasons for this have been offered over the years. In Robinson's original formulation of enlargements [20] a nontrivial familiarity with model theory and logic was required. This was noted as a barrier for most mathematicians. In [21], a set theoretic formulation of enlargements was offered which eliminated the need to know much model theory or logic. One only needed ability to understand bounded formulas, saturation, and a transfer principle. The popularity of this formulation for enlargements has been enduring. Still, it has always required a bit of on-the-job training to get acustomed to enlargements and use them freely. The standard copy correspondence $* : X \to X'$ entailed notational luggage as well as the split feeling of doing mathematics in two places at once. For this and other reasons mentioned earlier, the dream of global "non"standard set theory arose. In each of the proposed vehicles IST, NS_1, NS_2, NST, and $SNST$ the standard copy correspondence is suppressed and a single, albeit split leveled (e.g., standard/internal/external) universe is offered in which to work. At this point, as noted earlier, a clear jump from the ordinary mathematician's sense of ontology has occurred.

It is my thesis that this jump in mathematical ontology has always been inherent in "non"standard mathematics, that Robinson's standard copy corresponence merely disguised this jump at the cost of splitting mathematical geography[1], that the jump felt in the subsequent "non"standard set theories has remained a truncated jump, that the full jump is into a far reaching realm of ontological *relativity*, and that this fundamental ontological relativity inherent in "non"standard mathematics is the real reason for the wider mathematical community's difficulty in taking up the cause.

In designing the vehicle EST, I have deliberately ignored the needs of practitioners and sought instead to decisively illustrate the full implications of this relativistic mathematical ontology. The EST cosmos is in continual explosion. Its universes A are merely frozen frames of reference. Within these frames, familiar mathematics can be observed, but with the slightest slip along the chute of time — i.e., a passage to any further enlargement $A \subseteq A'$ — the usual develops unusualness. The universe A will have its clear version ω_A of the natural numbers, but when viewed in a later universe A', this ω_A will be seen to have picked up new, unusual elements which are infinite. Of course A' has *its* version $\omega_{A'}$ of the natural numbers, but this only means the cycle of slipping "reality" is poised to repeat itself again.

Appealing to my role as a visitor in these realms, I choose to end this work by leaving the question of how the system EST may be used as an *open problem*. It can be conjectured that we still don't know the full power of Robinson's original blending within set theory of elementary extensions, saturation and external constructions. This appears to ultimately involve a radical departure from conventional mathematical ontology. It may take a future generation of mathematicians to take to this inherent relativity and work with ease within it. To the young mathematician and to the mathematician who is young-at-heart, the system EST is offered as a proving ground. Rather than give a few half hearted examples of EST in use, which would be both unconvincing and also leave the impression of closing the case, it is preferable to consider the real use of a system such as EST as being an open ended project for the long term. One can speculate that the persons who makes significant progress in this endeavor will have remembered Leibniz well. A key to understanding EST's relativity will involve invention of efficient notation to gracefully display the insight. A rereading of Chapter 12 shows an initial taste of relativity where the model theoretic symbolism $A \models \varphi$ is *temporarily* useful for discussing truth within different frames of reference. A Leibniz would ponder deeper improvements.

---

[1]See [2] for a more recent proposal of a global vehicle for "non"standard mathematics which reintroduces Robinson-style standard copy correspondences, now suitably globalized. The working universe is ZF^-C^+ + superuniversality and hence is ontologically near-standard.

References

1. P. Aczel, *Non-well-founded sets*, Stanford, 1988, CSLI Lecture Notes, number 14, Center for the Study of Language and Information.
2. D. Ballard and K. Hrbáček, *Standard foundations for nonstandard analysis*, J. Symbolic Logic **57** (1992), 741–748.
3. J. Barwise and J. Etchemendy, *The liar: An essay in truth and circularity*, Oxford University Press, New York, 1987.
4. M. Boffa, *Forcing et negation de l'axiome de fondement*, Classe des Sciences, Memoires: Collection in-8^0, ser 7, vol. 40, Academie Royale de Belgique, 1972, no. 7.
5. C. C. Chang and H. J. Keisler, *Model theory*, third ed., North-Holland Publishing Company, Amsterdam, 1990.
6. W. C. Davidon, *Near-standard analysis*, 1986, preprint.
7. F. Diener and K. D. Stroyan, *Syntactical methods in infinitesmal analysis*, Nonstandard Analysis and Its Applications (Nigel Cutland, ed.), London Mathematical Society Student Texts 10, London Mathematical Society, 1988, pp. 258–281.
8. U. Felgner, *Comparison of axioms of local and universal choice*, Fund. Math. **71** (1971), 43–62.
9. P. Fletcher, *Nonstandard set theory*, J. Symbolic Logic **54** (1989), 1000–1008.
10. C. W. Henson and H. Jerome Keisler, *On the strength of nonstandard analysis*, J. Symbolic Logic **51** (1986), 377–386.
11. K. Hrbáček, *Axiomatic foundations for nonstandard analysis*, Fund. Math. **93** (1978), 1–19.
12. A. E. Hurd and P. A. Loeb, *An introduction to nonstandard real analysis*, Academic Press, Orlando, 1985.
13. T. J. Jech, *Set theory*, Academic Press, New York, 1978.
14. T. Kawai, *Nonstandard analysis by axiomatic method*, Southeast Asian Conference on Logic (C. T. Chong and M. J. Wicks, eds.), North-Holland, 1983, Elsevier Science Publishers B. V., pp. 55–76.
15. J. L. Kelley, *General topology*, D. Van Nostrand Company, Princeton, 1955.
16. W. A. J. Luxemburg, *A general theory of monads*, Applications of Model Theory to Algebra, Analysis and Probability (W. A. J. Luxemburg, ed.),

Holt, Rinehart and Winston, New York, 1969, pp. 18–86.

17. T. Lindstrøm, *An invitation to nonstandard analysis*, Nonstandard Analysis and Its Applications (Nigel Cutland, ed.), London Mathematical Society Student Texts 10, London Mathematical Society, 1988, pp. 1–105.

18. E. Nelson, *Internal set theory: A new approach to nonstandard analysis*, Bull. Amer. Math. Soc. **83** (1977), 1165–1198.

19. _____, *The syntax of nonstandard analysis*, Annals of Pure and Applied Logic **38** (1988), 123–134.

20. A. Robinson, *Non-standard analysis*, North-Holland, Amsterdam, 1966.

21. A. Robinson and E. Zakon, *A set-theoretical characterization of enlargements*, Applications of Model Theory to Algebra, Analysis and Probability (W. A. J. Luxemburg, ed.), Holt, Rinehart and Winston, New York, 1969, pp. 109–122.

22. K. Stroyan and J. M. Bayod, *Foundations of infinitesmal stochastic analysis*, North-Holland Publishing Company, Amsterdam, 1986.

23. K. Stroyan and W.A.J. Luxemburg, *Introduction to the theory of infinitesmals*, Academic Press, New York, 1976.

24. P. Vopěnka and P. Hájek, *The theory of semisets*, North-Holland Publishing Company, Amsterdam, 1972.

25. A. N. Whitehead and B. Russell, *Principia mathematica*, second ed., Cambridge, England, 1925-7.

Index

Symbols

Recent Titles in This Series

(Continued from the front of this publication)

(See the AMS catalog for earlier titles)